DO CATS NEED SHRINKS?

PETER NEVILLE

DO CATS NEED SHRINKS?

SIDGWICK & JACKSON

LONDON

Acknowledgements

With many thanks to Claire for all her patience, love and skill, and to Bullet, Scribble, Wandsworth, Nimrod and Thumper, the cats in my life so far. Thanks too to all those veterinary surgeons who refer cases to me, and especially to Ian Hughes and partners, Keith Butt, Geoff Little and partners and Professor Keen and Tim Gruffydd-Jones and colleagues who provide such excellent facilities and expertise to support this pet shrink in his works. Thanks also to my long suffering parents who have to tell their friends what their son does for a living, and finally to cats and their devotees everywhere for all their inspiration.

First published in 1990 by
Sidgwick & Jackson Limited
1 Tavistock Chambers
Bloomsbury Way
London WC1A 2SG

ISBN 0 283 99980 2

Phototypeset by Claire Bessant,
Bessant Neville Partnership
Salisbury, Wiltshire

Printed in the UK by Mackays of Chatham plc

To Claire, it's time I popped the question

DO CATS NEED SHRINKS?

Contents

Problems Problems

Introduction

'If man could be crossed with the cat it would improve man, but it would deteriorate the cat'   Mark Twain

'Mr Neville? I'm sorry to 'phone you so early on a Sunday, but it's my cat... no, it's my husband really. He says the cat's got to go. It's done it once too often and he says it's either the cat or him... and you're my last resort.'

So begins another Sunday at 7.30 a.m. Gently, I edge my own usually hyperactive and hyper-vocal, but now thankfully asleep, Siamese cat Scribble and her ex-feral friend, Bullet, to one side and crawl out of bed into my office to grab the other phone.

'Now, Mrs...'

'Fisher.'

'Let's work through this logically so that I can get a better picture. First, what sort of cat do you have, what sex is it and how old is it?'

'George is a girl. We thought she was a boy but she isn't. She's a one-and-a- half year old moggy. We've one other cat and they seem to get along okay but she just won't stop doing it!'

'What is she doing, Mrs Fisher?'

'She pees everywhere. But it's not *peeing* peeing because she's standing up. She also does puddles behind the sofa and my husband says he's had enough.'

'How long has George been doing this?'

'About three months, ever since next door's cat came in and had a fight with her in the kitchen. Oh please say you can help, I couldn't bear to have her ...well, you know, at the vet's...'

1

It's a typical case. A normal, much loved cat has suffered the trauma of invasion by a local rival right in the safe core of her territory. The incident has long passed but the consequences on poor George's behaviour continue. It is a relatively common case of nervous urination in a safe place indoors rather than going outside where she is too vulnerable when squatting. Spraying the house is a desperate attempt to keep a hold on those resources she cannot leave – food, love and shelter. The much loved cat has quickly become unloved by one member of the family and relations between the owners are clearly strained. I may be the last resort. The fate of the cat and perhaps even the marriage may hinge on my successful treatment of the problem.

Firstly I ask Mrs Fisher to talk to her vet to make sure that there are no medical reasons behind George's behaviour, and to ask him or her to refer the case to me. Then I'm off to town to make the necessary house call at the Fishers' and take a closer look at the problem. Two hours later and we've worked out why George is behaving in the way she is and compiled a programme of treatment involving modification of the home environment, controlled exposure to sensitive areas and restructuring of the relationship between owners, especially Mr Fisher, and George. Within a couple of weeks at most the problem should be resolved.

As I was leaving the Fishers' to enjoy the rest of my Sunday, my client casually remarked, 'George has been munching her way through the back of the bedroom chair since she was a kitten. Is there anything you can do about it?'

Not content with being a sprayer and a nervous toileter, George is about to form part of my continuing research into the phenomenon of fabric-eating by cats. Now that *is* worth a bit more of my Sunday.

On rare quiet days I often wonder how it came about that I should be a 'pet shrink'. Having passed through the usual early ambitions of footballer and explorer, and then developing my lifelong interest in animals by studying biology at university, I still could never have predicted that I would become a professional consultant in pet behaviour. Nor could I have known that my Sunday mornings would be spent treating spraying, toileting, fabric-eating cats. That it should happen to me is perhaps at least understandable, given my attachment to animals, but that our society should have a place for a consultant in feline behaviour is undoubtedly surprising to many, not least those of my father's generation. Thirty years ago, a cat was an

2

animal that scarcely warranted feeding, let alone veterinary care or treatment from a cat psychologist.

So times must have changed and our relationship with our cats must have altered, for me to do what I do. This book is about the types of problem that I treat and the people and cats that I meet. Above all this is a book for those who, like me, are incurably fascinated by cats, and an insight into the more unusual aspects of their behaviour and their relationship with us.

1
Cats, Dogs and People

Cats have rarely been much use. Aside from the often overstated mouse- and rat-catching role, cats have not followed the pathway of domestication based on training to task followed by the horse or the dog, nor the pathway of providing us with food (with certain eastern countries excepted) as have farm animals such as the cow, pig or chicken. Nonetheless cats have always been with us – worshipped in Ancient Egypt, yet sacrificed as accomplices to witchcraft in Europe in the Middle Ages. Rejected and doted on, the cat is now even more popular as a pet in the United States than man's traditional best friend, the dog. The same is set to happen in the United Kingdom and probably elsewhere in northern Europe by the mid 1990s.

Despite this popularity we have achieved little success in altering the basic design of the cat compared with the way in which we have manipulated dog genes to produce such breeds as the Chihuahua, Great Dane, Mexican Hairless Dog and the dreadlocked Hungarian Puli. Instead it is human beings who have changed: physically we are now on average larger than at any other point in history and socially we are changing even faster. The cat, a perfectly evolved top predator, is completely suited to come with us into our modern, self-created technological niches. People like cats because they are clean in small modern apartments. They are self-determining and self-amusing and loved especially because they are affectionate to us in an increasingly formal and business-like world that affords little opportunity for emotional expression. Children are growing up ever earlier these days and starting to reject the nurturing, protective behaviour of their parents far sooner than they did twenty years ago.

In contrast the cat is there to love and will always — well nearly always — respond to our desire to initiate affection.

In a world where security means a written contract and emotional expression is often seen as weakness, the friendly cat is one of our only real-life outlets for uninhibited expression. How reassuring that on our return home the cat will be pleased to see us, purr and rub round our legs in an unconditional, uninhibited display of affection. And how reassuring that we can feel free to respond to that display without our fellow man belittling us or perceiving our expression of emotion as an opportunity to take advantage while our defences are down. Being affectionate with your cat or dog is socially acceptable. And nowadays people need that as they never have before.

Why not dogs?

'To his dog, every man is Napoleon, hence the constant popularity of dogs.' Aldous Huxley

Our association with the dog goes back approximately 10,000 years. The relationship was primarily based on the dog's willingness to assimilate into our social group, in return for the benefits of food, shelter and protection. We behaved as a dominant pack leader towards him and he was prepared to adopt the subordinate role at the bottom of the hierarchy in our social system. The benefit to us was that the dog would, and of course still does, share in the role of protecting our common interests: at first fending off the wolves from our livestock, and then by guarding our homes and shared territory. Even today a dog's bark is seen by many insurance companies as such a deterrent to robbery that they offer cheaper home insurance policies to dog owners.

Dogs were often 'kept' for their speed and skill in hunting, so improving our survival prospects. Nowadays their physical skills of scent following are cleverly applied to drug and explosive detection work and their ability to be trained is invaluable to the blind, the disabled and to the police in apprehending the ever-increasing numbers of villains. But while it is true that the number of dogs being kept in the United Kingdom is rising at a rate of about ten per cent per year, fewer and fewer of us can afford the luxury of a dog simply as a companion, as 'man's best friend'. As in earlier times, the dog must once again do something to earn his keep and justify our patronage. The popularity of the larger, tougher guarding breeds to protect us and our property perhaps indicates that the modern role of dogs in our cities is increasingly returning to the 'utilitarian' as it was all those thousands of years ago. It's just that we – not the wolves – are now our own worst enemies.

Protection in the form of a hefty Rottweiler or barking German Shepherd Dog may be necessary for survival in some areas of our cities not sufficiently wealthy to afford entry phones at the door or full-time surveillance systems. But we should alreadyall feel uneasy at encouraging dogs to be aggressive and worry about how controlled that aggression is. We also worry about our environment, and the use of parks and open spaces as dog toilets. When this extends to pavements, the outcry is even louder – there is nothing worse than stepping in it.

Add to that fears of enteric worm and other diseases transmissible from dog to man via their faeces and we have a pet under pressure in our crowded cities from the 'green' point of view. However good they are at protecting us, dogs are getting more and more unpopular with the non-dog-owning public and calls for tighter legislation over ownership, licensing and physical control are heard frequently in the press and from politicians.

Cats too can infect us. They can carry enteric worms, and deposit their faeces outdoors in soil, or less desirably, in children's sand pit play areas where worm larvae may be inadvertently ingested. While the infection of man by the roundworm *Toxocara cati* is as yet unproven, compared with the established infection with *Toxocara canis*, the risk of exposure to a protozoan parasite causing toxoplasmosis from a feline is higher. Over 50 per cent of the population will be infected at some time, yet few of us ever develop symptoms. The real risks are to women who become infected when pregnant, causing abortion or damage to the baby. Ensuring that cats are wormed regularly and that extra care is taken when cleaning litter trays or gardening will minimise the chance of infection. The main risk is from meat which is not cooked properly – so keep off the extra rare when pregnant.

In cities it is the wealthier, younger section of our society, the yuppie and the dinky (double income no kids yet), who lead the way in the keeping of cats. They can often afford those non-canine security systems for their property and so need a pet for more emotional reasons. Cats are favoured by socio-economic groups AB and C1 while dogs remain most popular with the C2 and DE groups according to the annual survey of pet ownership trends published by the Pet Food Manufacturers' Association. Cats are kept by the busy younger age group but also in owner-occupied houses where the head of the household, male or female, is in full-time employment. Thankfully not all pet owners are alike or easily categorizable and not all well-off people live in cities. Many people decide to keep no pets at all or, like this pet psychologist, have both cats and dogs (just in case you thought I was being anti-dog in any way).

The well-respected Yale-based human/animal bond philosopher Stephen Kellert proposes that all pet keeping falls into four categories. Forgiving, if you will, the Americanese, there are those in our society whom he describes as 'negativistic', who dislike or are afraid of animals and actively avoid them. Many of this group have pets 'thrust upon them' later in life and find they have been missing out on something rather special. However, as few 'negativists' will be reading this, I'll pass on to the next category which is 'utilitarian'. Utilitarians' main concern for animals centres on their practical or material value. Older members of our society predominate in this class, having been brought up in harsher times when every mouth to be fed had to contribute to the family in some way.

While many would-be utilitarians in the West have no need to keep pets now, younger people in the 18-35 age bracket can enjoy being 'humanistic' or even 'moralistic' because present-day wealth allows them the indulgence. Humanists are interested in and have a strong affection for individual animals, and most pet owners nowadays fall into this category. Moralists enjoy the further luxury of being concerned with the rights of animals and are opposed to all forms of cruelty from a moral and often campaigning stance. The moralist attitude is almost incompatible with the utilitarian ideal and, of course, unthinkable in the Third World or even amongst the worst-off in the Western world.

It is perhaps a paradox that many elderly utilitarians are now benefitting from programmes of pet-facilitated therapy designed to improve the lives of our institutionalized elderly, physically disabled and mentally ill. Cats and dogs are increasingly being taken into hospitals and even penal institutions to help rehabilitate or simply provide some undiscriminating love and affection. We all know of the relaxation benefits in lowering heart rate and blood pressure derived from even the simple action of stroking a cat or dog and perhaps we have designed a society where we have come to depend on that facility.

For modern man, life is getting too unpredictable to keep a dog. Dogs have to be walked, exercised and fed at set times. They can't be left alone at home for too long or accidents happen. Nowadays both partners in a relationship usually have to work to pay for city-style mortgages and so are out all day. Late or variable working hours and plenty of places to go for fun afterwards would just mean a miserable life for the dog. Yet after a hard day and evening at the office, an after-hours' business meeting in the wine-bar, trip to the theatre and late-night drive home, making the obligatory call to relations and abandoned friends on the car phone, modern man needs his pet. He has to stay human somehow! Increasingly the pet of choice will be the cat.

The Director of the Companion Animal Research Group at Cambridge, Dr James Serpell, argues that, in the past few decades, not only have we shifted our relationship with animals – except those large guard dogs – away from the 'utilitarian' approach of many of our parents, but also our increasing 'humanism' has allowed us to accept and develop strong emotional attachments to our pets that would previously have been regarded as unhealthy or odd. An inevitable consequence of this close relationship and the lack of human company in big cities (you're never so alone as when you're alone in a city surrounded by people) is the viewing of our pets as people, the process of (deep breath) anthropomorphasization.

While small cage pets such as hamsters are also increasing in popularity they are simply not highly enough developed for the type of emotional involvement we now seek. No matter how aesthetic and calming to watch (I hope your dentist has a tank in his or her waiting room) you can't really have an exchange of emotions with a tankful of fish, so again we think cat.

Cats do well as apartment creatures despite being so highly adapted as territorial predators. That's why there are over 55 million in the United States where over eighty per cent of people live in cities, and over 6.7 million in the United Kingdom, where more and more of us live in cities or suburban areas. Easy to keep, adaptable and happy to enjoy what we offer without depending on us, the cat was never so perfectly suited to us. Cat haters, of which there are still many (few are ambivalent towards cats, you either love 'em or hate 'em) often despise the cat's very independence and desire to be fed, sheltered and loved on its own terms while remaining aloof. They may prefer the slavish devotion of the dog. In biological terms the cat's relationship with us could be described as parasitic, compared with the symbiosis of our relations with the dog, but that would be to place no importance in economic or survival terms on keeping cats, when we now need them more than ever before for our emotional well-being.

Cats have rarely been of much use, yet they have stuck with us throughout thousands of years. And, as that relationship evolves, new demands and responses start to emerge and the resultant sanity of cat and owner is what this book is all about.

2
The Relationship

The earliest known forerunners of the present-day domestic cat appeared on earth about 13 million years ago, with the larger forms such as lions and tigers not arriving until about 10 million years later. Inbetween times the cat was so successful at the top of the food chain that it spread around the world to all parts except Australasia, Antarctica, Madagascar and the smaller islands.

The most likely ancestor of our modern pet, *Felis catus*, is the reddish to grey/brown African wild cat, *Felis lybica*, which is slightly larger and has longer legs, a leaner body and a long thin tail. This species tames readily and the earliest records of man's association with any cat date from 2600BC in Ancient Egypt which was part of the African wild cat's territory. Evidence of feline domestication can be found in Egypt from 1600BC onwards where there are many tomb paintings depicting the cat's role in home and family life. By this time the Egyptians had already been domesticating other species for over 1000 years so perhaps the late arrival of the cat in Egyptian homes is due to social changes. The more advanced a civilization becomes, the better it cares for its weak, and therefore the more popular the practice of keeping pets becomes. It is also possible that the cat underwent a genetic change which made it more docile and tolerant of man, enabling it to come into his house from the outside. Early handling of young born indoors facilitated taming, and thereafter followed a deliberate selection of friendlier, responsive individuals.

This is largely supposition, of course, deduced from what little archaeological evidence there is available, but while the African wild cat was able to evolve into a domesticated species, by contrast the pure tabby European wild cat, *Felis sylvestris*, has never been tamed. As the African wild cat spread or was transported as a pet by man, it interbred with its European cousin to produce the original tabby markings of what we now know as the domestic cat.

Domestication itself implies a genetic disposition for docility which man is able to exploit and then deliberately select in breeding, to make his animal companions non-aggressive, less fearful and easier to handle. Clearly the modern farm species were originally selected for production, but docility was also essential if they were to be enclosed and bred efficiently and safely. Dogs too were selected for docility and function and it was only after thousands of years of producing tame dogs that we really started to manipulate the dog's genetic compliment for appearance and to produce the 450-plus breeds we know today. Most dogs still fulfil basic guarding roles for us but a few are kept for specific tasks such as hunting or herding sheep. They are with us as companions and in many ways have been so altered by us in the process of domestication as to have little chance of surviving without us. There are plenty of feral (domestic gone wild) dogs throughout the world, from New York to Liverpool to the high regions of Italy, but these do not survive long unless able to group with others and reform a successful hunting/scavenging unit as their ancestors did. Certainly the most successful feral dogs are the medium-to-large size mongrels with short-to-medium length coats, as dogs of the artificially maintained pedigree strains soon interbreed if they survive at all away from our direct care.

Domestication is therefore something of a compromise for many species, but less so for the cat. The loose association with man from earliest times when the cat was encouraged as a rodent controller around our settlements but not assimilated into the human family as part of the 'pack', has ensured that its nature has not been fully compromised by our efforts. Since 'domestication', cats have been allowed into our homes but they remain free to come and go as they please, so ensuring maintenance of their self-determining behaviour patterns. Of course, we have selected in a general way for tameness but, having evolved as a solo predator, the cat has no need of us for actual survival and hence no need to accept a subordinate position in our family for the benefits of food or shelter. As an opportunist the cat takes advantage of the food we offer and appreciates the shelter, using our house as a lair, but always retains the ability to fend for itself. The process of domestication never really touched that successful evolution as a hunter and all cats (perhaps with the exception of the squash-faced Persians) remain proficient and active predators given the opportunity or necessity. Even the best-fed cats still bring home birds and mice.

There are huge numbers of feral cats surviving and breeding to maintain populations, so they do not rely on immigration from stray or abandoned pets. Ferals are found in all corners of the world from the Galapagos Islands (where they have become a serious threat to the unique fauna) to the hotels of the Mediterranean and Indian Ocean coasts, the famous cats of the Colloseum in Rome, the streets of London and all cities in east and west, as well as on many farms. The domestic species reverted to a wild lifestyle seems to do far better than any of the original wild species, such as the Scottish wild cat, which are mostly in decline due to persecution by man and loss of their habitat. This is probably due to their willingness to stay close to man for ease of finding food. Some feral cats, such as those populations arising from cats abandoned on uninhabited islands by passing ships, do survive solely by hunting; others exist largely by scavenging from rubbish bins or from hand-outs provided by kind-hearted people.

The domestic species of cat is therefore not totally dependent upon us. Insead it survives extremely well on a range of feeding strategies when nearly all other species we have taken on are now too altered to survive without us. The reason they stay is because we make life easy and all that domestication has done is to select those cats that are better at learning how to get the best from us by displays of affection and relaxed companionship. But this is to paint too simple a picture of the cat's side of that special relationship with us. Uncompromised perhaps, independent certainly, but the huge bond we are able to establish with such a highly developed creature is not just a case of detente or mutual acceptance. Instead we must look to the developmental biology of the cat to discover where we fit into the cat's perception and why we are more than just tolerant cohabitees.

For the dog we are the pack leader, a Napoleon who makes decisions, orders life, protects the den and other pack-mates and allows access to food. This relationship is inconceivable with a non-pack animal. To a cat we are its mother, long after it would have become independent and, in fact, for the whole of its life. The only time in a cat's life when it would ordinarily be dependent on other members of its species is as a young kitten when it first needs its mother for milk and then to bring home more solid food when it is weaned. By this stage the kitten is already showing marked independence through exploratory behaviour and is having progressively less contact with its littermates. At only twelve weeks after birth it is able to fend for itself and is a fairly proficient hunter.

By contrast the twelve-week-old dog is very much a juvenile and still learning how it should behave within the rank- order system of the dog pack. It must learn to wait for its share of the spoils of hunting and that older, high-ranking dogs take precedence in selecting favoured shelter and walkways, and have the right of passage. Taking a puppy from its mother after weaning at six to twelve weeks and placing it in a human pack is simply to redivert the usual assimilation into the dog group. It is noticeable that dogs homed early after weaning (at about six weeks), which are handled frequently and, after vaccination, exposed to as wide a variety of people, other dogs and environments as possible, are far less likely to present behaviour problems when growing up or as adults. Lack of sociability with other dogs or people due to nervousness or aggression is largely avoidable given the right sort of early exposure. Keep a puppy isolated during that crucial period of six to eighteen weeks old and it will inevitably be incompetent later.

The same principles apply generally to the socialization of kittens. Frequent early handling and homing soon after weaning at six to eight weeks is most likely to produce tractable, playful and friendly young cats, even if they are still destined to be territorial and not very sociable with other local cats. After all, none of the small wild cat species is particularly sociable in the wild. Yet if we take feral kittens born in the wild and unhandled before about eight weeks of age, we can quickly tame them and make them into pets which are completely indistinguishable from home-reared kittens. However, take a six to eight week old Scottish wild cat and it is beyond taming; instead it follows the rejection patterns and developmental independence pathway we have all probably experienced in trying to raise young wild rabbits, squirrels or other mammals. While they depend on us and are young, they are friendly, accepting us as a mother substitute. Once nutritionally independent and entering adolescence, they reject us and move on to establish their own territory away from the 'maternal' home.

Even more interesting than the prospect of taming feral kittens is the fact that many adult feral cats which have never been handled or had any contact with man can also be tamed into perfectly friendly and responsive pets. It takes a little longer than with feral kittens and may require a concerted effort and the use of intensive socializing techniques to achieve it, but it is being done by cat rescue groups all over the country. However, the same efforts with a true wild cat barely achieve recognition behaviour between man and cat.

The feral cat is able to live in a wide variety of social systems depending on the availability of food. Where food is scarce feral cats are likely to be solitary and intolerant of other cats but as food supplies increase their capacity to be sociable rises. It is probably not worth risking injury in a fight when there is enough for everyone and so large numbers of cats may be found together, all apparently tolerant of each other. While many groups comprise related animals in an extended family arrangement, others are made up of unrelated feral cats that sometimes develop a complicated system of social organization. They rarely, if ever, hunt together or share prey, yet they will eat together peacefully and one mother may suckle another's kittens and share in their protection. It seems that domestic cat society is highly variable and extremely elastic – another feature that has surely contributed to its ability to assimilate into so many different human lifestyles. Perhaps it was sheer chance that the African wild cat possessed or developed the genes which allowed it to be tamed by the Egyptians and perhaps we owe the success of the pet cat today to that chance, for without it all cats would possibly still be shy, distant creatures and declining in numbers like most of the small wild cat species.

So our modern domestic cat is equipped to accept us and live with us. But this can only be achieved because we fulfil the role of its mother and mimic the only time in its life when the cat is truly a social animal. We not only provide food when the young adult cat would be expected by its mother to find its own, but we also allow the cat to demand physical attention, warmth and affection from us as only its mother could provide. We offer the same security in our laps as the kitten knew when suckling or lying next to its mother. Often the cat responds with purring and affectionate rubbing, and also by kneading our laps as it used to knead its mother's nipples to stimulate milk flow. Some particularly affectionate or closely bonded individuals even salivate and dribble at these times in anticipation of a milk feed. Dribbling perhaps excepted, we love this infantile display of affection, and cannot resist the urge to stroke our cat and be affectionate in return, while sitting with him on our laps or cradled as a baby in our arms. Rarely is the relationship one of 'child substitution' for us, yet

14

relaxed, warm and protecting, we all play the role of a cat's mother throughout its life and, as a result, relax and feel better ourselves.

At other times we play with the cat in simulating hunting by offering moving targets of ping-pong balls and string to chase as its mother would have enticed him with her tail to teach him to stalk, chase, pounce and master other predatory skills. As the kitten was weaned, its mother would have brought back half-killed small prey for it to learn to catch and despatch and thus further learn the hunting game. Instead we provide full meals, already despatched, which simply need eating, so enhancing the providing mother image. Small wonder cats are so affectionate at meal times. Lastly, we provide a safe lair – our house – where the cat is protected by four walls (and a whole family of mothers) from enemies and competition with other cats. This is also the place where those meals and affection are provided so it is always worth returning to, unless of course there is better service down the road.

Our cats need us and we encourage the dependence constantly, but the maternal relationship is as much a function of cat behaviour towards us as ours is to them and even when totally indulged as true 'child substitutes' they retain the ability to grow up quickly and survive well without us. We build up enormous emotional attachments to our cats – there is little option when playing a maternal role – but more than this, we are starting to *need* to care for something in our modern and increasingly emotionally vacuous society, and the cat is the perfect receiver for the release of our maternal and loving emotions.

Cats and culture

Our bond with the cat has developed from loose attachment for the benefit of rodent control, through tolerance of a non-competitive self-reliant creature, to dependence, but throughout the ages the cat has also been admired simply for its grace of movement and beauty of form. In the closer modern relationship that admiration has become more pronounced. Poets and artists have exploited feline attributes since Egyptian times and the languid shape of the cat was a major theme in the Art Nouveau of the 1920s in the West.

Nowadays we are all at least armchair naturalists. The modern media of television, cinema and highly advanced photographic techniques have brought the diversity and complexity of nature into our homes and enthralled us all. As a result we are more educated and no longer regard the predators at the top of the food chains – such as lions, pike and eagles – as ruthless, harmful killers. We can now understand their role in keeping nature fit by feeding on the weak, sick and old and as a result can view the predatorial side of our pet cat, if not with pleasure, then at least with understanding. We still marvel at his graceful ease of movement and sharp reflexes but we also understand the evolutionary benefit and functional adaptation of the various features of the hunter. Highly developed senses and efficient weapons of claws and teeth are all part of the divine plan. This opportunity to have the tooth and claw of nature contrasting with its absolute beauty and lying benign by our artificial fireside in the city is yet one more reason for the increasing popularity of cats in human society. Urban tigers one minute, babes in arms the next – a perfect contrast.

The law

We house, feed, admire, mother and love our pet cats. We claim ownership, yet our lack of success in actually training them to perform set tasks or breeding them to fulfil certain purposes has resulted in a recognized lack of ability to be responsible in legal terms for their actions. A cat cannot trespass and you cannot be held responsible for damages if he causes an accident, as a dog owner could. Legally you are only liable for injury your cat may cause to another cat or person if it can be proved that you knew your cat to be aggressive and wantonly released him. So we enjoy a very pleasant, low-risk relationship with one of the world's most finely adapted killing machines! The cat as a piece of property is, however, afforded legal protection, and persons injuring or killing it are liable to make financial reparations for the damage. And we can reclaim a stolen cat as long as six years after the theft, even if the present owner is unaware of the cat's background. It seems that the cat has managed to achieve the best of all worlds once again and even the law has had to accommodate its nature.

3
The Support Industry for Cats

Whether we love our cats for their wild side or their affectionate indoor character there is no denying that, for the most part, as a society we look after them extremely well. In the United Kingdom we keep 6.7 million cats in 4.5 million households – so thirty-five per cent of those have more than one cat (you may be a 'twitch' – a two income, two cat household). In the United Kingdom we spent over £2 billion on all our pets in 1988 – as one daily newspaper wrote, 'about 58 times the gross national product of The Gambia and 23 times the World Bank's agricultural grant to beleaguired Ethiopia in 1987'. This is morally questionable in such terms, but a glance at the huge industries now centred around the pet cat alone indicates its importance in creating employment and generating income in the United Kingdom.

The leading brand of canned cat food occupies the largest shelf facing for a single grocery item for either man or beast in our supermarkets, and the lion's share of a market selling 524,500 tonnes of wet, dry and semi-moist cat food, which sells at a retail price of £443 million per year. In Europe as a whole, where just under one in two households keep at least one pet, 25 million cats consume 1,070,000 tonnes of cat food annually. Most of this is derived from the meat or fish industry and uses only material which would otherwise be wasted. None of the companies which make up the UK Pet Food Manufacturers' Association, from whose 'Profile 1989' annual report this data is presented, use any pony, horse, whale or kangaroo meat or by-products in their pet foods, so your cat can eat freely in the knowledge that he is not offending British culture with regard to equines or the conservation movement. Cat biscuits and treats

17

command a £3.5 million market at the till — less than a tenth of the dog treat market but rising rapidly as we feel we need to reward cats.

The cat litter industry is worth nearly £40 million, despite the fact that most cats use the great outdoors. The litter industry is expanding rapidly with some of the larger chemical and mining corporations moving in. That largest single grocery product manufacturer has also recently launched its own litter brand to present a complete package for food and waste. The litter manufacturers stand to gain much with the increasing numbers of cats being kept permanently indoors in towns and cities, where our cats have a greater need for their own private facilities.

We spent about £500 million at the vets in 1988 on all our pets, so cat health is also big business for veterinary drug companies. They vie for the veterinary surgeon's favour for vaccines or treatments in an increasingly similar high pressure manner to that used in the human medical market. We have little choice at our vet's as to which vaccine is injected into our cat to protect him for the next year against the unpleasant or fatal viruses of influenza or enteritis, and are probably not concerned or very knowledgeable about it anyway. We happily leave it to our vet and let him or her be bombarded by the various vaccine manufacturers and make the best choice on our behalf.

The days of the James Herriot-style vet are numbered, as agriculture becomes more intensive and the range of knowledge so vast that even now most practices are either 'small animal' or 'large animal'. Few are mixed as in Herriot's day. Our vets now represent the peak of academic achievement, and aspiring animal lovers are unlikely to gain one of those much coveted places at one of our six veterinary schools without three A grades in the sciences at A-level. This is largely due to the popularity of the profession, much of it fuelled by *All Creatures Great and Small*, but also because the five or six year veterinary medicine course is highly demanding intellectually and requires a range of skills, not just an empathy with animals. It is interesting to note that it is far easier to enter medical school and train to become a human doctor.

There are approximately 10,000 vets in the country, though not all in general practice: some work for the government or are in industry or research. There are about 2700 actual practices to choose from and, despite the difficulty of entering veterinary school, it is still a largely vocational profession. You can rely on your vet for a sympathetic ear and a high level of compassion that is all too often missing from our

18

own health facilities. It's not that doctors don't care, it's just that they are overworked and don't often have time for those important touches within a state health system. The days of the family GP look set to change with profit incentives as I write. Doubtless there will be an increased burden on counsellors and psychologists to do the listening part of the doctor's old brief, at either the patient's or the state's expense. The National Health Service unfortunately mitigates against the vet in practice because few of us are aware of the true cost of treatment when we are ill. We simply go to the doctor or into hospital and are made better, without a bill. When our cats are sick we trot along to the vet, expect the same equipment to be available to diagnose the problem and the same competent treatment, to make them better. Then we get a bill and complain.

Vets have a tough time of it in many cases simply because we have precious little idea of the cost of health care, let alone an appreciation of the skills involved, which include a range of surgical skills for different species that would require a group of specialist consultant surgeons in the human field. So next time your cat is unwell or due for his annual booster vaccinations, take a fresh look at your vet and be grateful that the bill isn't half what it would be elsewhere in Europe or America. But if you are still worried by the prospect of even a fairly large bill you should consider insuring your cat or kitten against the costs of unforeseen veterinary treatment. This can cost as little as £30 per year and, although routine vaccination, neutering and other non-essential surgery is excluded, whenever your cat is unwell you will be able to tell your vet to carry out whatever treatment is possible, irrespective of costs. Many pet insurance policies are also prepared to cover some or all of the costs of treating behaviour problems, which is comforting for owners and myself alike.

Feline medicine has made major advances over the past ten to fifteen years as our demand to keep our cats healthy has risen. In this country the Royal College of Veterinary Surgeons prohibits vets from establishing practices to see only one species, but feline-only clinics in Australia and the USA are becoming more common and, like the university referral departments at UK veterinary schools, act as important collectors of information as a result of the huge numbers of cases seen. At the University of Bristol Veterinary School a special interest in feline medicine is maintained and respected worldwide. A specialist feline charity, the Feline Advisory Bureau, raises money almost solely to fund their 'scholar's' post within the Department of

Veterinary Medicine. The scholar is a qualified veterinary surgeon with a special interest in cats, and each is resident within the department for one year to see cases referred by vets in practice and investigate feline medical topics of special interest. The post has been such a success that the Cats' Protection League, another cat welfare organization, has set up a similar scheme in the same department, and now a major producer of small animal vaccines has established such a post at Liverpool University Veterinary School Small Animal Hospital.

So if you live within striking distance of Bristol and your cat is ill with a condition not responding to your vet's treatment, or one which is little described or requiring special equipment for treatment, your own vet can refer you to the university for special attention. Similar referral facilities exist at the other veterinary schools. It all adds up to the best possible service for when the cat is ill. No wonder that cats are now living longer and it is not uncommon to hear of twenty year olds, when even ten years ago a fifteen-year-old was quite notable. Incidentally, the world record is thirty six years and one day, achieved by a British moggy from Devon called 'Puss'!

So technologically, we have a rapidly advancing country, part of the self-help cat-loving community of Europe, and a society that is not only keeping more cats but is also generating huge support industries in all areas from food to private health insurance.

4
Do Cats Need Shrinks?

Cats represent extremely good value in terms of the pleasure we get from them, as each cat has two quite distinct characters.

Indoors, the cat is affectionate, placid except for odd bursts of activity, and manages his life around us without getting in the way. Outdoors he snaps back into his wild alter-ego and becomes the hunter once again. While we may disapprove of his successes, we still stand and stare enthralled at his stalking skills and his tenacity at defending his patch of garden territory. Home and hunting ground defended, birds put to flight and mice to ground, he pops back home and reverts to his kittenhood at the drop of a hat.

The vast majority of cats live out their nine lives in this very attractive style, which couldn't have been better designed to suit their every whim. A few highly prized pedigree cats, those at risk near busy roads and increasing numbers in our cities are denied the opportunity to express their wild component and are kept indoors for all their lives. While this may seem a pity in that the owners never get to see the uncompromised predator in their cat, the animal itself rarely seems bothered. Provided the cat is taken in as a kitten and never allowed to experience the outdoor life, it can adapt perfectly well to apartment life. A case of 'what you've never had, you don't miss' applies to most, although it would be unthinkable to try to confine a cat to the indoor life if it was used to patrolling a territory outdoors. Frustrations would understandably develop and alternative, and perhaps unwanted, behaviours arise such as pacing, late-night crying to be let out, and over-reactions to minor events, such as chasing and biting passing legs.

At last, a mention of the possibility that our cat's behaviour could sometimes be less than perfect. The solution may be obvious in this case – either take the risk and give your cat back its established freedom, or rehome it so that it can start to enjoy life again. But for other problems with less obvious causes, treatment may prove difficult. For the cat that soils your carpet, sprays your kitchen worktops or runs and hides himself away as soon as the doorbell rings, the quick solution of rejection by re-homing or asking the vet to end it all may be the only prospect after you've tried all the best advice of friends, cat breeders and the local vet.

But most of us are far too involved in that maternal relationship for such callous expediency with a creature that, at its worst, is perfectly well behaved for ninety-nine per cent of the time, and one on which we have come to depend for more than just companionship. Small wonder that most owners of cats with behaviour problems put up with them.

Owners often blame themselves for their cat's behaviour problems and feel that their cat is simply reacting to a lack of love or failure to provide the right type of home environment. They may worry that they are inadequate owners, rather than let the blame fall on the 'innocent victim'. As a result the emotional compromise of loving a 'rotter' who continually abuses all that love and care can place enormous strain on owners and their families and, ultimately, far outweigh the joys of owning the cat in the first place. Owners of house-soiling or indoor spraying cats invariably reduce their human social contacts because they are embarrassed to let people see their 'unclean homes' and so their dependence on the cat for company rises still further and any desire to dispose of him falls. The circle becomes vicious and the owners more and more frustrated, until eventually family relations may start to suffer and they are forced into taking some sort of action.

But just what constitutes a problem is highly variable. For some, a single mistake on the carpet when their cat is 'caught short' is sufficiently damaging to their enjoyment of keeping him to cause rejection. Most put up with the same event as something to be expected from time to time, and accept that what with fur and footprints, owning a cat or a dog inevitably means accepting a lower standard of home hygiene. Many others put up with far worse problems without turning a hair, almost regarding their cat's unhygienic lack of toilet training as normal. The definition of a problem is a highly personal matter based on owners' individual tolerance levels and their relationship with their pets.

A behaviour problem is often only acknowledged when the pains of coping with a cat start to exceed the joy it brings. If the owner of a cat finds any aspect of its behaviour or their relationship problematic, it ill behoves anyone to tell them that they are worrying unnecessarily. But just where owners can turn to for help to tackle any problems is rather limited. First they may be unwilling to reveal that they have a problem with their pet because of the guilt factor and the necessity of letting it be known that all is not well and that, by implication, their pet is less than happy with them. They may be particularly unwilling to disclose house-soiling or indoor spraying problems because of the reflection on the standards of cleanliness in their own home. But if the problem simply isn't responding to the owner's efforts to resolve it, then they must 'come out' and admit that they are sufficiently attached to their cat to suffer in putting up with his problems.

Owners usually turn first for advice to the relative safety of friends or relations who also have cats, and then to their cat's breeder or a well known local cat fanatic. The vast majority of problems are resolved by such help or, if not, at the next port of call – the vet's. Experience often tells, particularly with house-training problems in young cats, but there is a small minority of problems which are either so rare as not to have been experienced before by even breeders or vets, or which are so complicated that there is no easy or straightforward solution. Then there are other persistent problems, including many indoor sprayers, which either don't respond to treatment or only temporarily, and relationship problems where the cat simply isn't responding to the owner's desires or expectations, perhaps because of nervousness or low threshold aggression. Tackling these types of problems, be they of feline behaviour, relationship incongruities or, as is often the case, a combination of both, takes time. Time to look at all the factors that may be influencing the cat's behaviour, from the structure of its home environment to its relations with each member of the family and its friends or rivals from along the street. Time to look closely at the cat's development and early upbringing in the true psychologist's style of going back to childhood, and looking at its general character and disposition aside from the problem. No two cats are exactly alike, no two owners and no two problems, and so a very broad approach is required to identify likely causes of each problem, and to develop a personal treatment plan with each owner. It takes time, a good background in feline and general animal behaviour and, most of all, a caring approach in dealing with a very important and sensitive aspect of people's lives. Lastly, you need to be able to motivate owners to tackle the problem when they may be nearing the end of their own patience.

It sounds like a blueprint for a consultant in feline behaviour with just a touch of human psychology and counselling skills, and it is. Do cats need shrinks? Perhaps they themselves don't and neither do most of their owners. What they do sometimes need is a feline behaviourist with the time and patience to understand why the cat is doing what it's doing and when, and how to modify it within the framework of the cat's relationship with its owners and the geography of its home territory. The relationship we have with our cats can also need adjusting from time to time, and owners can be given a little advice on how to restructure that relationship to bring about changes for the better in him and alleviate certain types of problem behaviour. The cat's behaviour can need only a little improvement, or sometimes a lot, for owner and cat to appreciate life and each other to the full once again, or sometimes even for the first time at all.

And the more we enjoy keeping cats and the more we find emotional release in them, the more concerned we will become when things do not run smoothly. As in so many other areas, we increasingly look for expert help from a detached professional viewpoint to help us iron out the difficulties when our own efforts fail to resolve those problems. The days of punishing or rejecting a cat for the crime of having a behaviour problem are over because help is now available and given the right approach, those problems can usually be resolved. If the strength of relationship between man and cat wasn't sufficiently strong even ten years ago to warrant a consultant in feline behaviour, it is fast becoming so now.

In Practice

My practice runs strictly on referral from veterinary surgeons and I work almost exclusively alongside veterinary surgeons at their premises. Obviously a cat's behaviour, like ours, will change when it is ill so it is important that the cat is first examined by a veterinary surgeon to eliminate this possibility before I see the case. Other cases, such as cats which mutilate themselves, are best seen by a combination of vet and behaviourist as the behaviour may arise from influences in both 'camps' and treatment involves a combined approach. But ordinarily, I see cases in private with the owners and their cats for approximately one to two hours at two veterinary hospitals in the north-west of England (where I see dog behaviour referrals as well), the University of Bristol Veterinary School, and at a veterinary practice in south west London. I also make a few house calls for owners who have difficulty travelling to those centres. During each consultation I record details about the nature of the problem, and the cat's and owner's lifestyles. Having discussed the problem I then try to work out a programme of treatment for the owners to carry out at home. After the consultation I forward a written report to them and a copy is sent to the referring veterinary surgeon, together with any suggestions for drug treatment such as the prescription of sedative or hormone treatments that I feel would assist treatment. The owners stay in contact with me by phone and we reassess treatment after two to three weeks. Only rarely will I need to see a case twice.

I always count myself as very lucky because not only do I get to work with my favourite animals professionally, but I also get to meet the nicest people. Pet owners are more sociable than non-pet owners, and whatever their pet's problems, cat owners especially are always pleasant, sensitive and interesting. More importantly, they usually have a splendid sense of humour and this is reflected in the following chapters where we will be looking broadly at the types of cases I treat and referring to past cases, typical and atypical. Only the names have been changed to preserve client confidentiality and to protect the innocent.

PROBLEMS PROBLEMS

5
Nervousness,Phobias,Insecurity

Dear Mr Neville,

I have two female spayed cats, not related, one four and one five years old. The four-year-old, Sammi, is an extrovert, totally confident and very affectionate with everyone including the dog. Shelley, by contrast, is timid, runs and hides from every noise and creeps around in fear all the time. They both go out but Shelley never goes far and runs indoors at the slightest disturbance. In fact she's really only at ease on my lap. Should I worry? Is there anything I can do to make her enjoy life more? It seems such a pity when Sammi is so happy.

Yours sincerely

Monica McFarlane

The aim of being alive, the evolutionists tell us, is to live long enough to procreate and ensure the survival of our genes. The ultimate aim of this process is unknown and almost too magnificent to contemplate. While we can trace the development of many life forms on the earth over millions of years through studies of fossils and traces left by species long since extinct, no-one as yet has been able to create even the simplest species from a bottle of the right type of chemicals, nor have they been able to show why the survival of the fittest of each species, or even each species itself, is important. Not being a philosopher or an expert in divine purpose, I should perhaps confine myself to how each cat strives to survive and pass on its genetic compliment to the next generation.

Human intervention, of course, denies the prospect of reproduction in most cases by neutering the cat, yet the will to survive obviously persists. The cat, like every other animal, is born with the capacity to respond to challenge and protect itself from the life-threatening dangers of attack by predators and environmental risks, such as fire. The responses to such challenges are easily distinguishable from the cat's normal (most often expressed) behaviour patterns. The startle reaction is genetically programmed and even young kittens are seen to arch their back, erect their fur and flatten their ears when alarmed by a loud noise or sudden unfamiliar happening. During the early weeks of life, a kitten, like many other mammals, will run to mum for protection immediately after being startled, if she hasn't already intervened. Gradually the kitten habituates to commonly encountered noises or environmental stimuli, especially if they are never actually followed by anything harmful or physically painful.

In short, the kitten grows up constantly learning which stimuli are potentially threatening and which are not, yet without ever losing the capacity to demonstrate the initial startle response. In the average upbringing each kitten is challenged by a huge number of environmental changes and social encounters, first with its littermates and mother, and later with other cats. It learns to adapt and take most things in its stride. Once learned, its competence at dealing with a given stimulus without showing a fearful reaction is maintained by occasional exposure to it. Repetition of exposure, maternal teaching and the development of exploratory behaviour, which propels the kitten into new environments, all help it learn to deal with novelty and use those innate patterns of behaviour to deal with anything threatening. Monica McFarlane's Sammi is a typical competent, confident cat which has gone through the feline school of life and, suitably manipulated by us, has become the ideal pet.

For others, nervous responses such as seeking to avoid any challenge by remaining in a dark corner or behind a solid protecting barrier like a sofa, or low threshold, over-quick flight reactions when

27

experiencing slight changes to the environment, ensure that the cat never learns to cope. The anxious cat is easily identified by his crouching gait, low carriage of tail, and slow low movements towards a sheltered spot under a table or in a corner. Once as secure as he can make himself, he may then face the world, still hunched and trying not to draw attention to himself, pupils dilated as a response to adrenaline, a hormone released at such times to prepare his muscles for running away. Alternatively he may adopt the ostrich method of avoiding challenge. Having found a secure bolt-hole, he then avoids facing the challenge altogether and lies still, hoping for it to disappear. Some even learn to hide themselves under the bedclothes when anxious, similar to that famous 'head in the sand' method of making worries go away. Whichever reaction the cat chooses, it is clearly distressed, and it is distressing for the owner to observe as well, especially if the cat reacts in such a manner in response to harmless, everyday events and never gets to grips with life. The signs of his distress are clear, yet it is difficult to comfort the cat at such times without contributing further to his anxiety. At the very time he is trying to avoid challenges, we draw attention to him, and he may never learn that our efforts are designed to comfort him. If we push too hard, he may even lash out as a last line of defence when there is nowhere else to run and the threshold of his nervousness falls ever lower as a result. One such cat is Shelley, the opposite in the competence ratings to Sammi, and clearly not enjoying life to the full.

Unhabituated responses

Kittens brought up by inexperienced or incompetent mothers will be less competent themselves at dealing with change, as indeed will those relatively unexposed to challenge during the first weeks of life. Their survival chances in a wild world would not usually be so good. There is enormous individual variation in each kitten's threshold of reactions to different stimuli, even among individuals in a litter apparently exposed to the same events from birth. While young animals always habituate to stimuli faster than adults, the smaller or inherently less robust individual may hold back from its littermates in advancing to investigate a new toy because it has learnt that it may get caught up or bowled over in the rush. It will avoid this risk by waiting until last, or not advancing to investigate at all. As a result it will experience one less event than the others and be less competent by comparison. Faced with the same toy later, the other kittens will already have learnt to overcome any reluctance to investigate and leap straight in to play or ignore it. By contrast, the inexperienced kitten may show an initial startle reaction to the toy when finally encountered or, worse, never dare to experience it at all.

Perhaps Shelley is an example of the adult result of the reluctant and incompetent kitten. Such general nervousness is termed unhabituated anxiety – the cat simply received an inadequate range of experiences as a kitten to enable it to cope with the flow of challenges as an adult. Controlled exposure techniques can be designed to allow Shelley to experience what we and Sammi would regard as commonplace and non-threatening events in the home and even outdoors. In short we can provide fresh opportunity for Shelley to learn to cope with all those happenings. However, in such cases it is rare to produce a totally competent, 'normal' cat as there are always certain stimuli which can only be habituated to as a kitten because of subsequent negative experiences or the failure to have positive ones. For example, kittens brought up with puppies, or even adult dogs, rarely show nervous reactions when dogs approach them as adults to play. By contrast the adult cat which has no experience of dogs is unlikely ever to be relaxed about such advances and usually retreats quickly. Some older cats can learn to live with a dog late in life, but it takes an inordinate amount of patience and most only get as far as tolerating dogs and rarely will any social interaction or play develop.

When treating any nervous animal, progress can only proceed as far as its present capacity to learn allows in the light of its sum total of experiences. Improvements can be made with cats like Shelley by providing the cat with a new den (a kittening pen is ideal) which is placed in the main activity room of the house. Normal family life goes on as usual around the cat, which is protected from physical danger by the pen. Most importantly, the cat is also prevented from running away and avoiding the challenges of changes in personnel, movement, noises on the television or movement of furniture going on around it. So it has to face up to them and start to interpret what is happening. The cat is essentially in a womb, warm and given all life-support systems of food, water, and, unique to wombs, a litter tray, but in a position to assimilate and habituate to stimuli that previously induced a fearful, avoiding reaction. The same techniques can even

29

be used outdoors with certain agoraphobic cats (see later).

Severe cases of unhabituated nervousness can also be helped with sedative treatment. Yes, cats with certain types of problem can be given valium, as can their long-suffering owners. In treating most nervous conditions it is important that the sedative is gradually withdrawn as the cat's competence to deal with 'life' improves with controlled exposure, so that its new competence does not become dependent on the drug. In short, the drug is a 'vehicle' for the learning process so that the cat gets the best out of its capacity to learn without being overshadowed by an unnecessarily pronounced initial fearful reaction. In some cases a low maintenance dose of sedative may be required long term, or even permanently, for the cat to be more confident but, fortunately, the majority of even severe cases need only a short course to accompany the exposure side of treatment. While I am no expert in alternative medicine, certain types of homeopathic or herbal treatment have also been reported to serve as effective alternatives to sedatives and are increasingly being prescribed by veterinary surgeons in certain cases. But, as with conventional veterinary medicine, it is vital that advice be sought only from a properly accountable practitioner who is a member of the British Association of Homoeopathic Veterinary Surgeons.

Unhabituated nervous cats will usually worsen if left untreated. The majority of such cases referred to me have concerned cats aged three years and more. They are often described by their owners as having been shy or withdrawn from the day they arrived and their tolerance of the owner's lifestyle and home goings-on has steadily worsened. The threshold of their flight reaction falls to the point where almost any noise or quick movement is enough to make them run to a bolt-hole, and stay there long after the 'problem' has passed. Other typical signs include increasingly secretive, withdrawn behaviour by the cat, which comes to avoid all change, and take an ever increasing time to come out into the open areas of the house when the owner returns after a period away. The cat seeks dark, quiet corners for shelter and spends long periods there, and may even only venture out for food under cover of darkness.

The picture is often a sorry one and highly distressing for the owner to watch, yet the signs of a worsening condition have been evident for some time. The longer the cat is left untreated, the harder it is to reverse the trend of withdrawal and to restore the cat to its original level of competence and beyond to greater things.

More rarely the unhabituated nervous cat may cling to its owner constantly for security and show marked and generalized nervous reactions when parted. Over-dependence is particularly difficult to treat for the owner as well as the cat because it involves a loosening of that bond, but it should be treated for the sake of the cat's emotional state when away from its owner, if nothing else.

Work by behaviourists around the world has clearly established the importance of frequent handling of kittens between the ages of two and seven weeks to habituate them to human contact and lifestyle, and make them into tolerant and friendly pets. Essentially we overcome any initial startle reaction by flooding the kitten with approach and contact, which it soon learns is not threatening. It finds that such handling is even enjoyable, being similar to the contact provided by its mother. It also learns that our homes are safe, and that being in an open area in our home gets you noticed and more likely to receive some of that comforting maternal affection. The same is largely true of puppies, and by the time most of our pets are weaned and becoming socially aware, they are long over any fear of humans . . . usually.

Loss of habituated responses

Dear Mr Neville,

I read about you in a women's magazine and wondered if you could help with my cat. I should say that I have had cats all my life and am glad to say that none of them has ever sprayed, soiled in the headphones of the hi-fi or eaten my underwear. But Topper is now nervous of visitors and always runs away when anyone calls. He used to be very tolerant and even friendly and he's still okay with me and my husband, but I feel so sorry for him being afraid to stay in his own home when my friends only want to be nice to him. As soon as the bell goes, he's off and won't return until after they've gone. Is there anything we can do? Help!

Yours hopefully

Joyce Hayes,

Topper is a prime example of the cat which learned to cope with the arrival of strange people in the heart of his territory, but then steadily lost the ability. I see many such cases where the cat is often otherwise untroubled by changes within the house, but it completely loses the earlier habituation to the stimulus of the arrival and presence of its owners' guests.

The converse of this problem is the frequently observed pastime of confident cats blessed with a sense of humour. They leap upon the lap of any visitor who dislikes or, even more unfortunately, is allergic to cats as soon as they sit down. The cat seems to enjoy the discomfort of the poor person who is naturally unwilling to offend their host by turfing their cat off, and instead shift or sniff uncomfortably while the cat dominates proceedings.

But Topper's nervousness of visitors may not necessarily have been caused by lack of suitable experience with enough different people when young. It could be due to a single unfortunate experience with a particularly noisy or unkind guest which frightened him, and taught him to avoid all risk of repetition in the future by running away early. Such single traumatic incidents also lie behind the development of many specific phobias in dogs and people, but I have yet to see a case of a phobic cat, other than agoraphobia where the cat is afraid of open spaces indoors or more usually, is afraid to go outdoors. Common canine phobias include pronounced fear of wind, rain and sound, and I have even seen cases of dogs which are telephone box and light bulb phobic! Nervousness of any type can perhaps be described as more general or less specific than a genuine phobia, but the principles of treatment are based on desensitization, with controlled exposure of the stimulus in a diluted manner that does not invoke the fearful or phobic response in the patient. The stimulus intensity is steadily increased and the patient habituates to it as a learning process.

Cats are natural avoiders of potential conflict as a survival mechanism and have highly developed senses to enable them to detect danger and respond quickly with flight if necessary. That they do not wait to investigate an apparently dangerous, stimulus such as a car backfiring, has helped to ensure their survival as a solo creature that cannot rely on fellows to protect it if caught napping. The difficulty in treating some cases of nervousness lies in determining what is 'normal' and the cat which is judged to be 'nervous,' and whose quality of life is affected as a result.

Clearly Topper is not enjoying the potential benefits of meeting his owner's friends and regards them as undesirable threats to be avoided by running away. This is a perfectly normal reaction designed to ensue the cat's survival; it's just that there is no reason for the cat to perceive that its survival is threatened. The more guests the owner has, the less she sees of Topper, and the less he regards home as a secure base. The end result is a worsening relationship with the cat, which may at all other times be thoroughly rewarding, or a change in the family's social life, because they stop inviting friends round for fear of upsetting the cat. Either way there is a problem to be resolved, and in most such cases this is perfectly possible.

However, treatment is rarely simple and does not involve hanging on to a panicking cat while trying to press him into the arms of even the most understanding cat-loving guest. Throwing in at the deep end doesn't always teach you to swim – it can cause drowning. In cases such as Topper's, such action would be guaranteed to enhance his fear of visitors by adding the prospect of restriction by his owner to his apprehension, quite apart from any pain experienced from the owners frantic efforts to keep a tight hold. Fortunately most owners only try this approach once and call me while their injuries are still fresh. Such cats usually calm down quite quickly once at a safe distance (the flight distance), groom themselves and then get ready to run when they hear footsteps approaching the door, not even waiting for the doorbell to ring.

Treating Topper involves the same sort of controlled exposure techniques as when treating Shelley, but targetted more directly at overcoming the specific problem of accepting visitors. The first aim of treatment is to block Topper's attempts to escape or avoid exposure to the challenge. His success in doing so, although protecting him from the danger he perceives, also precludes any possibility of him learning to cope on his own. Instead Topper is denied the opportunity to avoid visitors either by being restrained on a leash if he is comfortable wearing a collar or harness, or by being kept in a travelling cat basket for short periods when he is to receive guests. This is placed in the area where guests are invited to relax – usually the living room – before they arrive. The more willing volunteers there are the better, although Topper should first receive 'guests' that he knows, such as members of the family. They ring the doorbell instead of using the key. Topper's first reaction is the usual one of alarm and an attempt to escape, but this is prevented by the basket. Then the 'guest' enters and Topper, seeing that it is only Joyce or her husband, quickly calms down. Repetition should cause Topper to start to associate the doorbell with non-threatening arrivals.

Later, less well tolerated guests can be asked to perform the same routine, entering Topper's room with an accepted member of the family and doing nothing more than sitting down some distance from him. It is essential that Topper grows used to their presence in gradual stages. Now his cage serves to protect him from the challenge he has avoided for so long, and he should settle quickly.

Slow progress at this stage can often be speeded up by quelling the cat's over-reactions with a little sedative treatment such as valium on prescription from the veterinary surgeon. However, as with cases of severe unhabituated anxiety, it is essential that the cat's tolerance does not become dependent on drugs, and a tapered dose prescription is usually employed. Initially the tolerance of guests may be totally drug-dependent for a few days but as he is exposed to more visitors, the drug is slowly withdrawn, so that his tolerance is increasingly learned and decreasingly drug-dependent. The drugs are simply a vehicle for exposing Topper to his problem. On or off drugs, with frequent exposure to as many different people as possible under the right conditions, Topper should perceive their arrival and occupancy of the core of his territory as being neutral. More importantly, they are able to remain inside his flight distance, the distance which defines his opportunity to escape in the same way that zebras maintain a flight distance from the lions on the African plains. It is a fundamental distance in defining chances of survival in nature. Those that misjudge the distance they need to effect escape from any sudden challenge through illness, age or incompetence get eaten first.

The next stage of treatment is a little more invasive. Now guests are asked to sit progressively closer to Topper's cage to habituate him further to their presence. This stage can only proceed as fast as Topper can tolerate and guests should certainly not attempt to touch him or even talk to him until he seems confident about their presence. Then we introduce the prospect of forcing the issue a little. Though it sounds a little unfair, Topper should be starved for twelve to twenty-four hours so that he is hungry when pressed into sharing space with his next visitor. The visitor, sitting close by Topper's cage rather than bending down over it which would alarm him, gently proffers a small titbit or tasty portion of a favourite food through the bars of the cage. Topper's positive appetite-stimulated reaction to the food is designed to equal or be in excess of the intensity of his nervous reaction and so help him overcome it. Again this stage of treatment can be assisted by certain drugs from the vet which increase the cat's appetite and therefore his incentive to come forward to guests. The drug of choice, which interestingly is used more generally as a contraceptive for cats and dogs, also has a sedative effect and so may be used from the early stages of treatment in severe cases. Food cements relations far quicker than gentle voices, though the visitor and owner should encourage proceedings by talking gently to the cat while offering food. Thereafter the cat should be fed frequent short meals for the length of the visitor's stay (or patience) and as many guests as possible, as well as the family, should take part in the process. This steadily spreads the cat's confidence around and helps him view all guests as potential providers of food and, later, affection.

The final stage of treatment is to dispense with the cage and restrain Topper on a harness or collar and lead when guests arrive and to have them offer food as before. Then the owner holds him as before and takes him towards a single known and accepted visitor. This should be done slowly so that Topper doesn't panic as he did before he learned that visitors could be nice. The advance should be slowed or halted if Topper starts to look alarmed or struggles. Once alongside, guests can gradually start to stroke him. This is most readily accepted if Topper is first stroked by Joyce, the guest slowly joining in. The process is complete when the owner stops and the cat is only being stroked by the guest. Holding Topper may not be possible for guests for some time yet, if at all, because this is an enclosing action denying escape and thus involving Topper's total confidence. He would only be acting in the same way as a great proportion of cats if he were to restrict this honour to his family only.

During all contact guests' hands should either approach initially from the side of the cat unseen, or very slowly from directly in front of Topper so that he can see and accept the advancing paw. The hand may be very much a paw to Topper and should be offered very gently indeed, bearing in mind that paws are a cat's main armoury and therefore likely to cause him to be apprehensive in the same way as he would be apprehensive of a swing from a cat's paw.

It may even help to approach Topper at the same level rather than intimidating him unneccessarily by bending over him. While the prospect of crawling along the floor to approach him may seem like carrying things to extremes, placing Topper – whether in the basket, on a lead or hand held – on a table and approaching face to face could be less threatening. Of course all volunteers must be safe and if there is any prospect at all of Topper lashing out defensively with a paw, then he should be in his basket when he receives guests for some time yet. Approaching face to face mimics the greeting behaviour of cats, and as our face shape is not too dissimilar from that of a cat with regard to forward facing eyes, it may be more readily tolerated than a big mit landing on top of Topper's head. Obviously all volunteers should avoid making loud noises or sudden movements as they approach him to avoid inducing the startle reaction and, equally, they should avoid staring at him as this could be seen as a challenging gesture similar to the staring 'battle of wills' observed in cats facing-off over territorial disputes. The non-territorial Topper would always run to avoid that type of confrontation with a visitor, so they should look below his eye-level or to one side as they approach him.

Progress with such cases is normally steady and, once improved, the cat rarely regresses provided he is regularly exposed to visitors. Essentially the methods of treatment are very similar to those with many other behaviour problems. The opportunity for success of unwanted behaviour or response is denied physically, the physiological reaction of fear or panic is muted by pharmacological means if necessary, and the animal made to face up to the stimuli which cause it to be nervous, in a slowly presented, controlled fashion. Gradually the patient learns first to accept, cope with and tolerate the stimulus, and ultimately perhaps even comes to enjoy it. The key to success is to reduce all variables so that everything is controlled and all possible reactions from the cat are expected, and to proceed in short frequent doses of exposure at a pace that the cat can take without inducing panic. When not being exposed to visitors, the cat should be allowed free range as before but should be placed in its basket periodically without being confronted by visitors so that he doesn't come to perceive it as an unpleasant restriction. Ideally he should be fed in it at normal mealtimes to maintain the idea of the basket having a variety of uses, some of which at least are good.

It usually works. However, children, loud, bad tempered or drunk guests or disabled visitors may still be alarming to Topper after he is apparently improved because of their unpredictability and less rational behaviour. Given that controlled exposure techniques may be difficult to organise, he should perhaps be protected from such unusual invasion by being put out or in another room before they arrive. This will preclude any possibility of regression to his old fearful ways

36

unless, of course, drunkenness is more the norm than sobriety, as was once the case of a cat I saw that had been rescued from a pub in Scotland. The cat, named Pernod, was very suspicious of people who were quiet and walked straight, but had no difficulty at all in coping with staggering loud drunks. All a question of early exposure, no doubt.

Agoraphobia

Dear Mr Neville,

My cat Mugs is agoraphobic. He is very happy indoors and used to be happy outdoors, but for the last four or five months he won't go out at all. I've tried pushing him but he just panics as we get near the back door and runs straight back inside if I push him out. If I shut the door he just screams by the back door until I let him in again. So now he has a litter tray in the kitchen and never goes out and I'm sure he's missing out on life as a result. Do you think he is suffering? Should I insist he goes out?

Yours sincerely

Jonathon Ringer

Agoraphobia is the fear of open spaces, which includes the outdoors, but can also mean open areas within a room. Fortunately this is quite a rare condition in cats and is the only true phobia I have ever encountered in the species. The condition can arise from a lack of early exposure to the outdoors, or exposure delayed too long and not coinciding with a kitten's exploratory phases of development after weaning. Sadly, from the behavioural development point of view, kittens cannot be allowed outdoors in safety until their vaccination course against unpleasant and often killer diseases is complete at about twelve to fourteen weeks of age. Indeed responsible pedigree cat breeders are forbidden by the Governing Council of the Cat Fancy to sell kittens until all vaccinations have been administered. Ideally kittens should be exposed to the outdoors as soon as they are weaned so that maximum competence at coping with the brave new world can develop, but clearly I would never recommend putting the kitten's health or life at risk to achieve this. Instead responsible breeders ensure that their kittens are exposed to as wide a variety of novelty, handling and challenge indoors in safety so that they all grow up with as wide a range of abilities to cope as possible, which will include exploratory behaviour patterns to help them cope with the outdoors as and when they are allowed access. Therefore my advice to prospective owners of a pedigree kitten is to seek out a breeder who is aware of the behavioural needs of kittens and who is not just raising money making cat bodies in an unstimulating pen at the end of the garden.

For such victims of callous breeders the prospect of being incompetent outdoors or even agoraphobic is a common development. It is perhaps surprising that so many kittens overcome such a lack of early stimulation to become well adjusted non-agoraphobic cats but, for a few, their unhabituated anxiety problems could so easily have been avoided.

The majority of the agoraphobic cats I treat are unwilling to go out or walk across the centre of a room because of a loss of confidence caused by a single traumatic incident, usually a fight with a highly territorial local rival. Trauma or uncontrolled exposure are the more usual causes behind the onset of most phobias. The most common such trauma in the development of agoraphobia is when a fight with a rival occurs in the cat's own house where it used to be completely safe. More often such fights occur outside and one suspects that such is the case with Mugs.

The risk of encountering the rival may be present every time he goes out and his reluctance to do so thus increases steadily. He may lose the ability to cope outdoors with even mild changes such as the sound of leaves rustling or the noise of a car, even if the other cat is not on the scene. Typically, the severity of the cat's reaction is pronounced with all exposures, even on apparently calm days when no other cat can be seen. But Mugs' agoraphobia may be due to other causes including major disturbance of access points to home base during the building of an extension or garage, a chance encounter with a stray dog in the garden or a near miss with a fast car. The consequences in all cases are usually quite marked and get progressively worse. The cat is clearly distressed if forced outdoors and a long, long way back from enjoying its earlier outdoor lifestyle.

Treatment for cats like Mugs is similar to other forms of nervousness and involves controlled exposure to the outdoors, perhaps using the carrying basket again or, better still, a large secure pen in which the cat can safely spend some portion of his day outdoors. Such treatment is usually best delayed until the cause of the problem is removed. This may mean waiting until building works are complete or even coming to an arrangement with the owner of the despotic rival about which cat is allowed out at which times, to avoid further conflict. Once achieved, Mugs can be put safely outdoors to re-learn that it is as safe as before the trauma and that every noise on the wind is not necessarily threatening. Jonathon should accompany him on the first few occasions and walk around the garden with Mugs on a lead so that he acts as a security bridge for him. Sometimes it helps to divide the cat's meals into frequent short rations and to move his feeding area to the pen and later just outside the back door to use appetite again as a distraction when the cat feels vulnerable.

The prognosis for treatment of cats such as Mugs is often very good, depending on the control that can be achieved over the cause of the initial problem, because we are re-establishing old habituation patterns rather than trying to teach them to the cat for the first time.

Husband phobia

Dear Mr Neville,

Our Persian cat, Fluffs, adores me, my children and even our rabbit, but hates my husband. The poor man has tried for months to get the cat to like him but to no avail. As soon as he enters the room Fluffs runs away. John has never laid a finger on him and only ever wanted to stroke him and be friends. Is there a way to get them together? Fluffs is just over a year old, and doesn't go out because we live on a busy road. He was neutered about six months ago but this made no difference to his dislike of my husband. It's really becoming a problem now because my husband gets upset that I devote extra attention to Fluffs to make up for what he can't get from him. Please help before we come to blows!

Yours truly

Karen Hayward

Dislike of husbands is more usually restricted to wives they tell me, and rarely if ever is described as a phobia! However, over the last year or so I have received a number of letters similar to that of the unfortunate Mrs Hayward. Some cats do bond very closely to one maternal figure but are usually tolerant of other members of the family without doting on them in the same way. For a cat to reject one particular individual while remaining friendly with every other creature in the house, including noisy kids and tasty prey items such as rabbits, is unusual and a real blow for the often innocent victim. As in this case, the victim is usually the husband. As in this case the husband has tried extremely hard to overcome the cat's dislike for him by trying to be ultra calm and affectionate. Sadly the efforts have failed and now other relationships in the family have begun to suffer because of the competition for affection from the cat, and then between the accepted family members and the person rejected. It may seem strange to begin to quarrel over a cat, but being rejected by any member of a social group is bound to hurt. If it happens to be a cat that you love dearly, treat very much as one of the family and spend much time and money on, then it is hard not to take the hurt personally or resent the success of others who perhaps expend less emotional effort in being nice to the wretched cat.

Why cats should be more likely to be nervous of the man of the house rather than any other member is difficult to say. With some cases the man is obviously short-tempered and prone to crashing around and shouting when upset. This is likely to upset the cat which will be unable to feel confident in his presence, even when he is relaxed and pleasant.

Yet with other cases the wife appears the more volatile and the husband is the calm, constant and loving influence in the household. Perhaps then we must consider the early experiences of the cat and investigate who took the major role in feeding, being affectionate and playing with him to understand whether the bonding process was a little unequally shared. Certainly for some husbands there is no obvious reason why their cat should dislike them. It may be that he simply isn't at home as often as the family and that his arrival sparks off much sudden family movement when he is greeted home from work, which the cat finds startling and so keeps out of his way.

I am told by a source close to home that one particular man may be likely to upset the resident cats at any time because of his general unpredictability. This man is apparently prone to great swings of moods being calm and pleasant one minute, then banging the table and making loud roars of disapproval on receiving some minor piece of unimportant news then, with emotions released, returning to his normal calm self. 'Small wonder,' says this informant, 'that cats might find men unpredictable or even undesirable and learn that keeping out of their way is a better policy than constantly trying to anticipate their moods.' But as both my cats like me, this cannot apply to the case she cites!

Treatment of husband-phobic cats has to be a family affair, and often this necessary corporate approach immediately restores the strained human relations to their former strength. The common aim of helping the cat draws the family together given the right explanations, even though the advice is not always too pleasant for the wife to carry out. She and other closely appreciated family members are advised to be far less available to the cat and to reject his demands for attention for a few weeks. Meals are divided up again into short feeds available exclusively from the husband, who is also the treat provider and the one who opens the door to let Fluffs in and out. Games are begun by the wife but are taken over by the husband and he is allowed the armchair closest to the fire or radiator (if he hasn't taken it for his own already) so as to be more attractive to the cat. He is advised to be more in control of his emotions and to present a more consistent appearance of calm and friendship for the cat.

The relationship usually builds more quickly if the husband gives up trying to pursue the unwilling cat and waits for the cat to need him, while providing a few incentives. And big chaps especially are advised to get down to floor level to mimic those feline greeting behaviours rather than lean down from on high, which can be frightening for the unsure cat. All fast movements and loud expletives are out, too. Progress may be quite slow in some cases but astoundingly quick in others, and for no apparent reason. In particularly slow cases I have even suggested that the husband stops wearing after-shave or 'stud' soaps, on the premise that many such 'masculine' toileteries are

41

scented with real or artificial male animal extracts such as musk oil and therefore present an unnatural challenge to the cat that we could never perceive.

The most rewarding letter I ever received after treating a husband-phobic cat explained briefly that the cat was much better and would sit on the man's knee to be stroked, and even occasionally purr, but best of all, the husband was a totally reformed character who had given up all signs of macho, short-tempered and dominating behaviour in the home for fear of undoing the progress and upsetting the cat. And for all that, they only got a small bill for the cat's treatment!

Nervous or phobic, insecure or lacking in confidence, the cat makes a good patient for the cat shrink because of its intelligence, adaptive learning capacity and highly developed individual survival instincts. Owners too are usually compassionate and more than willing to invest the necessary time and management to carry out treatment suggestions. Usually their efforts pay off and the cat's quality of life and their joys in owning him or her improve steadily.

6
Bond Problems

Dear Peter Neville,

I have the most wonderful cat in the world. His name is 'Baby', he is four months old and we love each other dearly. He only wants to be with me when I am at home and follows me constantly. He sits on the edge of the bath when I'm in it and waits outside the door when I'm in the loo. He'd follow me there if I let him. He's quite the most affectionate cat I've ever known and loves to be cuddled. He must think I'm his mum because he purrs loudly and sucks on my clothes as if they were nipples. I don't mind this at all except when he decides that clothes aren't good enough and crawls up to my neck and sucks it. Even though I have red rashes from his attentions I haven't the heart to put him down as I would hate for him to feel that I was rejecting his advances. Is there anything else I could do?'

Yours sincerely

Mrs Beverley Moore

Baby is an entirely appropriate name for this loving young cat who avoided the traumatic experiences of being isolated from its real mother by transferring all infantile dependence to his new human owner. It's simply a case of that maternal relationship we would have anyway with the cat, as described in the first chapters, being allowed

43

to continue to the same pitch with the owner. Although Baby has made the nutritional break of independence from his mother and eats solid food, not milk, the normal process of physical rejection and isolation from the caring, nurturing parent has not been completed due to the availability of a more than willing substitute. The result is that infantile behaviour and dependence on a maternal figure continues long after it should have been replaced by more adult patterns. The dependence on the mother for protection is maintained with the owner by following her everywhere so that Baby can reach her quickly should danger threaten. Normal exploratory behaviour and the development of independent decision making is markedly slowed, often to the point of incompetence or the development of nervous conditions when the cat cannot rely on its owner for security. The cat may even start to suffer from separation anxiety when the owner is away, a problem more usually encountered with the pack dog for which isolation is unnatural. For a creature which is largely an independent solo hunter when adult, separation anxieties ought never to occur. That cats such as Baby often 'fall to pieces' and hide themselves away in a dark corner when left alone is most definitely to be avoided.

However obvious the treatment of this case may be, Mrs Moore should be tackled gently. She is only doing what all cat owners do, enjoying that maternal relationship of being a provider of food, love and protection. It's just that she is probably being too conscientious with a young cat which is quite happy to continue being a kitten. The rewards of the kitten not becoming a cat are that it also retains much of the sporadic play behaviour we find so enchanting and demands our attention, which is in itself flattering, and enhances our affectionate feelings.

Detachment may be long overdue for the sake of Mrs Moore's neck as well as Baby's development but explaining the procedure without causing Mrs Moore to feel that she is rejecting her pet is a delicate affair. In this situation counselling skills may be far more important than any knowledge of cat behaviour. Mrs Moore should become far less available to Baby and affection should be initiated by her rather than by Baby. He should be prevented from following her and encouraged to spend more time developing new interests outdoors and given more opportunity to explore indoors. Mrs Moore's family should take over the caring duties of feeding and grooming Baby to help him spread his loyalties more, and be less focussed on 'mum'.

44

Most difficult of all, Mrs Moore has to offer far less affection and only in short doses, designed to mimic greeting and social bond-maintaining contact as found between friendly adult cats. Prolonged nursing type cuddling sessions should be stepped down steadily and never allowed to reach the point where Baby relaxes sufficiently to begin sucking of Mrs Moore's neck!

Baby will grow up quickly and become more independent, exploratory and competent to cope with life away from the apron strings. Mrs Moore may have to be helped to overcome feelings of guilt during the process, but my role is to help her justify the rejection and equate it in terms of allowing Baby to become a young boy without running the risk of losing his affection altogether. Perhaps a case of worrying too much, but a situation we can all unwittingly find ourselves in with even a well-adjusted adult cat if he happens to fall ill or get injured.

Intensive nursing

Modern veterinary treatment often allows us to nurse what were previously hopeless cases of injury or sickness in cats, with a good chance of recovery. While the latest drugs counter infection and the latest surgery resets bones and grafts skin, we as owners become the attentive nurse once the cat is allowed home. For days and weeks, and sometimes months, we tend our poor pets through all hours, feeding liquidised food through syringes and massaging limbs to keep them ready to function when our favourite cat starts to get about again. The cat can make an amazing recovery against all medical odds but afterwards he can be something of an emotional wreck. The problem is that over those weeks of nursing he has come to depend totally on his owner again for food, comfort, social contact and even to clean up when nature has called on the recumbent immobile cat. The psychological result is a recovered but over-attached cat who has become as dependent on its owner as when ill, or as a helpless kitten. This condition is fortunately not too common as the cat becomes increasingly independent again as it recovers, but for some owners the after effects of injury and illness are as tough and demanding to deal with as the nursing itself.

Detachment once again is required and in exactly similar fashion to the over-attached young cat, but again, strikes hard on the owner who has grown used to worrying about the cat's every need during its nursing. Good counselling skills are required to help the owner comprehend the need to ignore their pet's screams and wails for attention, their desire to be hand fed and constant efforts to follow them. But the owners don't always thank me for the basically unwelcome, if successful, advice. Anyone know of a consultant in feline behaviourist's behaviour specialising in rejection and anxiety problems?

Leaving home

Fortunately, there is the other side to the attachment coin, the side where my advice is to increase the volume or frequency of the contact and affection on all fronts to help a distressed cat cope with life or to enhance the bond with its owners. Baldrick is a cat who was feeling the pressure of his owners' hectic lifestyle and found life in their home increasingly unattractive. His owners worked shifts and he never knew who would be at home at any given time of the day or night. Feeding times were erratic or soul-less affairs of a cold plate of food, or just dry food left out by the owners as they passed through the house. A policeman's lot may not be a happy one, but this policeman and airport worker's cat's life was even less happy! Wild parties were frequent events when both the owners' days off coincided, and poor Baldrick had to cope with the contrast of having a quiet house to himself one day, and 3000 watts per channel all the next night.

Yet occasionally, one or other owner stayed at home all day, for the most part in bed, catching up on much needed rest, and then Baldrick was able to join the warm communal sleep heap. It was probably only these occasional opportunities to get together again that kept Baldrick around at all. Nonetheless, he steadily spent more and more time away fom home and grew distant from his owners. It was the summer of 1989, a long hot summer when many cats were happy to spend far longer outside, sleeping rough and hunting to feed themselves. I received a great number of calls that year from owners worried about their cat's long excursions away from home. Of course, as the autumn drew in and the weather became colder and wetter, their cats dutifully returned and spent more time at home, much to their owners' relief. It was a reminder for me of just how tenuous the relationship between man and cat can sometimes be, and just how readily the cat can return to his ancestral wild ways and survive very well without us.

Baldrick didn't go all the way, however. Instead he set up camp at the end of the garden under a lean-to next to the greenhouse. Sheltered, protected from draughts and warm on a pile of old newspapers, Baldrick learned to fend for himself and hunted for days on end in nearby fields. Yet he would also pop indoors from time to time to top up on the best quality free meals left by his owners some time previously. Once in a while he would come in to sleep a day away with one or other but the owners noticed a developing reluctance by Baldrick to be handled. It was taking an increasingly long time for him to recognise them if they happened to come home while he was indoors. His first reaction on their arrival was to run, and only sonmetimes could they call him back. But what really made the owners decide to act was the fact that Baldrick was starting to look like the grubby ill-kempt *Blackadder* television character who gave him his name. Baldrick the cat was a long haired moggy, who became more and more matted and flea infested as the summer wore on. His increasing reluctance to be handled, or even be near his owners unless asleep, was never so marked as when they tried to groom him....

The situation had gone too far for gradual modification of Baldrick's behaviour. I advised quick ruthless methods. First the catflap was set to allow Baldrick in but not out again. His favourite food was placed in a feral cat trap, a wonderfully ingenious device designed to catch unhandlable feral cats without hurting them, and keep them unhandled as they are transferred to an equally ingenious carrying basket with a movable side panel. In this Baldrick was taken to his vet, squeezed gently to the side of the basket by pulling the panel across and suitably sedated and then anaesthetised by the vet. Once unconscious, Baldrick was de-loused, de-flead, de-matted and thoroughly cleaned up and then put back in his basket to be taken home to recover. He was kept in a kittening pen in the front room for a week and only allowed out into that room when one of his owners was home. They adjusted their shifts to give him as much company as possible and we tried to establish regular feeding times for Baldrick around their rosters. If one could be at home within an hour of two feeding times, Baldrick would be fed by them. If not we employed another cunning piece of technology, the timed automatic feeder. Food is placed in a dish and the lid closed, timed to pop open and allow the cat access to his dinner at his usual time. I had one once for my cat

Bullet; unfortunately my faithful dog destroyed it in seconds to get at that most prized of foods.

The aim with Baldrick was to rebond him to his owners with a concerted effort over the week's confinement through frequent feeding, handling and playing, and as far as possible to re-associate his owners as providers of food at regular times. When not home the food would still only be available at those times. Hence the cat would always have a vested interest to return then, and the owners were usually guaranteed to find him in at those times. All meetings were thereafter to be effusive and, as expected, the more the owners invested in trying to react with Baldrick and be nice, the more he responded. Within a few short weeks Baldrick was part of the family again, though he is packed off to a neighbour's house for the night when all those strange airport workers and policemen arrive to play loud music, drink and not drive home!

The insecurities of old age

While Baldrick is an example of adapting a detaching cat to a strange modern lifestyle with no routine, sometimes it is we who must adapt to the cat's changing demands on us. This is likely to be true of all cats as they age, and with cats now living much longer, we are starting to observe similar behaviour changes to those in elderly people. Instead of the whole machine packing up, cats and people alike wear out a piece at a time. The slow wind-down or failure of an organ or joint brings inabilities to respond in the same way as we and our cats did when young. This in turn causes worry in us and an increasing awareness in cats and man of a failing ability to protect oneself or avoid potential hazard. For all cats, old age is a time for longer sleep periods and more time spent safe indoors, and for many it is a time to give up a little of that independent spirit and rely on one's benefactors for emotional protection as well as physical care.

> *Dear Mr Neville,*
>
> *Tiger and I have been together for over fifteen years and never have I suffered a single problem with her. Recently, however, she has begun to call pitifully at night. I get up and*

try to feed her or sometimes simply cuddle her for a while
and put her back on her favourite blanket by the radiator.
This stops the crying. My vet says she's as healthy as can
be expected of a cat her age and can find nothing wrong
with her. Why does she cry out? Is she in pain?

Yours, tired,

Anabelle Whitehouse

No Anabelle, Tiger isn't in pain. She's calling out because she needs the reassurance of your company. With many of the oriental breeds excepted, cats communicate with people and maintain their bonds of friendship through body language and the development of a communal smell through physical contact. Vocal communication is largely one-way from us to the cat, except during the first greeting or at feeding time when the cat will often issue forth a 'trill' of delight. Queens call loudly when in season to attract males but as most pets are spayed this is really only suffered by cat breeders! The most unforgettable feline voice is that of a constantly yelling Siamese queen in season, as many who thought it would be nice to let her have a litter will testify. Purring is rather different and full miaows and yells are restricted largely to the Siamese, Burmese, and other eastern races.

With increasing age, cats are not only less active and more dependent on us, they are also physically less able to get to us as required. Needing more support to feel secure they learn to call to attract our attention. Annoying perhaps, but the cat is most likely to feel insecure at night and periodically wake up and call out for re-assurance. We are aggrieved but always get up to see if the old boy is having trouble. Once reassured that we are around and protecting, the cat falls quiet and may be happy to go back to sleep. Reassured for the moment, he has also learned that vocal communication is the main method to be employed if feeling insecure and continues to do it for the rest of his days. At this stage the owner is probably already preparing smaller lower protein meals for their cat and allowing him to sleep in previously no-go areas because of his advancing years, so it's not usually too much of an imposition to allow the cat to sleep in the bedroom at night as well. He should be happy and comfortable, and at least if he does cry out and wake you up, you can reassure him without getting up!

Routines of feeding and contact, as with Baldrick, can also help to maintain a solid dependable relationship with the cat, and a secure bolt hole, warm, quiet and out of reach of dogs and children will be gladly accepted by the older cat. There are few places better than an old blanket on a shelf in the airing cupboard for this. Many cats teach their owners that trick along with that of the night-time calling.

7
Stress and Trauma

Stress is a little understood word that has so far been avoided in this book. Everyone knows what is meant by stress and virtually all of us suffer from it at some time in our lives, if not constantly, yet we know little about it. Cats, too, can suffer from stress and many of the problems I treat could be said to be reactions to stress of one sort or another. The interface between environment, behaviour and physiology is bared when we or our cats suffer from stress, and scientists are only slowly unravelling the complexities of it all.

Startle reactions and developmental learning help the cat cope with challenges and stay alive. Discretion is often the better part of valour for the cat, which will run away from danger quickly if the threat isn't discouraged by his startled arched-back posture of defence and warning hiss. More assertive individuals may actively repel living threats - a common example is that of the cat which has learnt to deal with over-friendly dogs by standing its ground and swiping that cold wet nose with a paw full of sharp dissuaders rather than running away. More nervous individuals may hide in safe corners or under bedclothes if unable to cope. Whichever method is adopted, the behaviour changes are designed to reduce the conflict of the moment.

Physiologically the cat is able to respond because adrenaline is released into the bloodstream from the adrenal gland, situated above the kidney. Adrenaline primes the cat for 'fight or flight' by causing the heart to pump faster and send more oxygenated blood to the muscles, while at the same time preparing the body to deal with increased carbon dioxide waste which will result from the activity of running away or fighting off the challenge.

Stress is a very common cause of behaviour changes in cats. Situations that can lead to stress include alterations in owner routine and lifestyle, retention in an unfamiliar environment, such as at the vet's surgery, or being locked in a cupboard accidentally. Overcrowding and prolonged exposure to worrying stimuli such as high frequency sound can induce stress, as can grief following the death of a companion, human or feline or, even occasionally, canine. Cats can also find it stressful if, when with other cats, their flight distance is not maintained; this is the distance that they need in order to be assured of escape should trouble start.

When our behavioural responses to challenge don't succeed in reducing it, the adrenaline continues to flow in order to keep the body prepared. Stress is often the result, and in human terms is manifested by irritability, tiredness, anxiety and depression. Heart disease and stomach ulcers can result from the very mechanism designed to help us cope with the risk of injury, if we are unable to avert the stressful challenge. Many of us adapt to stress and 'toughen' up both physically in coping with the rigours of, for example, sitting in a city traffic jam every day en route to work, and physiologically as our nervous systems evolve to be less sensitive to the adrenaline coursing through our bloodstream. This 'toughening up' is comparable to the learning habituation to stimuli described in chapter 5 (pages 29 - 30) in the physiological sense, and is a vital process in the adaptation to changing environment for any species. If the stress of moving from favoured forest shelter across an open plain was too much for a deer to the extent that it stayed in an old area with insufficient food rather than risk moving to a fresh area, then it would probably die out as a species quite quickly. That it can overcome the stress of moving across open land where it is temporarily more vulnerable, by physiologically adapting to its body's usual responses to such challenge, enables it to make the move and survive without suffering any harm from the stress itself.

Some people clearly enjoy being constantly in a state of readiness and, although stressed, do not suffer from any physical or psychological consequences. The difference between those who revel in such a state and those who are anxious, insomniac or get stomach ulcers seems to lie in the biochemistry of their respective nervous systems and their individual sensitivity at the neurotransmitter level, rather than in a failure to cope due to lack of exposure when young or through being over-loaded at first exposure, as when dealing with nervousness towards a particular finite challenge. Such an explanation, if true, at leasts helps us understand why everyone differs in their ability to cope with stress, and of course why each cat will also vary.

While there is no evidence of stressed cats developing stomach ulcers or heart disease, clearly many do exhibit behaviour changes and responses under stress which are similar to our own. The effectiveness of the immune system at combatting infection may also fall as the prevalence of certain types of white blood cell may drop dramatically with prolonged exposure to stress in man and cat. Low threshold aggression or irritability, depression manifested by secretive inactive behaviour and prolonged anxiety reactions all seem common. The appropriately named catatonia or 'playing dead' is seen in stressed cats and other animals and is probably designed to stop the chase response in a predator, prevent the attack and so help ensure survival. Indeed, feline responses to stress of any kind are similar to those of other mammals and generally fall into one of two categories: excited reactions or inhibited ones.

Less frequently reported in the cat and perhaps only observable in the artificial confines of the experimental psychologist's laboratory are convulsions, hysteric epilepsy, excessive salivation, panting, colic and even depigmentation of the skin (literally turning white with fear). The American veterinary surgeon and pet behaviourist Dr Bonnie Beaver also lists anorexia, pronounced sensitivity to touch, changes in taste preferences, diarrhoea, hair loss and even psychological neutering, an alarming prospect with which I am thankfully unfamiliar in UK cats as most of them have already been surgically neutered anyway. Additionally there are some peculiarly feline responses to certain types of stress, such as sweating from the footpads, urine spraying, and associative marking by scratching, urinating or defecating on socially or geographically significant areas.

Alleviation of any stressful influence when the cat's reactions have failed to achieve their aim is one of the main features of treatment of many behaviour problems referred to me. This often involves more than a little detective work in identifying those influences. They may be environmental, if the cat's security in the house is not defined and it therefore lives constantly under the stress of being competed with for the home resources of food and shelter. They may be social, if the cat does not get on with other cats that it has to share its house with, or with one or more of the people. But whatever the cause, treatment of stress in the cat is largely achieved by removing or modifying the stressful influences or controlling the cat's exposure to them physically so that it can 'toughen up'. Most cases involve a combination of both approaches and this treatment is usually far more successful than attempting to diminish the physiological reactions of the cat by using stress-relieving tranquilizer drugs which block the activity of the neurotransmitters. While the stressful responses may disappear for the duration of the prescription, if nothing is altered when the treatment is withdrawn, the cat will become stressed again and the behaviour problems soon recur. The temptation is simply to keep the cat on drugs constantly to prevent stress reactions but this is obviously only masking signs and not getting to grips with the real causes. However, used properly, tranquilizers can help in the treatment of stress in cats, and interestingly are usually prescribed at about three times the dosage neccessary to induce the same effects in man. But many cats respond quickly and well if the stressful influences can be identified and successfully manipulated without using tranquilizers long term.

Psychogenic dermatitis

Dear Mr Neville,

James is a very sensitive cat. He is now four and was neutered when about six months of age. I have no other cats because much as I would like one I don't think James would ever accept another one and would be jealous. He is generally rather a nervous little soul and although he enjoys going out, will often spend several days inside afterwards if he has a scrap with another cat. Sometimes he will not eat for a few days and simply hide himself away if I have friends to call. He can also be rather snooty at times and ignores me if I have been away for a while or won't respond to his demands for affection. While I love him enormously and can accept that he is a very sensitive cat, I am worried because when many of these problems coincide he will nibble at the base of his tail and up his back. Subsequently he has flaky skin which requires veterinary treatment. My vet feels that the problem is

54

psychological and so I am writing in the hope that you may be able to help.

Yours sincerely

Helena Bradley

Psychogenic or 'nervous' dermatitis is probably a more common feline reaction to stress than we realise. 'The problem with skin,' as a veterinary friend once said to me, 'is that so many things can cause the same problems and we often end up treating only the symptoms and hope the cause goes away on its own'. Now that there are veterinary surgeons with a special interest in dermatology, a few more of the causes of eczema and dermatitis are being unravelled, stress included. Diet sensitivity, particularly to preservatives, antioxidants and artificial flavour and colourings found in many canned diets, or lack of certain fatty acids, are suspected as causes of self-induced skin conditions by Mr David Shearer at Bristol University Veterinary School. Allergy (especially to fleas) and post-wound over-grooming can all set up skin irritations such as dermatitis and miliary eczema. I suspect that many stress-related conditions also arise which owners do not consider serious enough to warrant veterinary attention, and so they go unrecorded.

The normal adult cat spends over a third of its waking hours grooming to remove dander, parasites, matts and loose hair, which possibly also reduces the risk of parasite infestation. Grooming helps the cat to lose heat in hot weather through cooling evaporation from licking, and, most important of all, keeps the skin in good condition. It also helps enforce social bonds between friendly animals which groom each other. Indeed much of our relationship with the cat centres on his or her acceptance of our grooming-like activities of stroking and tickling. The tactile self-stimulation of grooming by licking, scratching or mild biting is also therapeutic in helping to relieve tension, as we too experience through scratching or combing ourselves when nervous.

A cat will often suddenly start to lick itself for no apparent reason. Perhaps it is responding to a minor irritation of misaligned fur or, as is often seen in the face of obvious stressful influences, as a diversionary activity which is more comforting than facing up to the problem. Cats such as James, which are sensitive to change or stress, or those

55

constantly challenged by unresolvable threats of one sort or another, may groom more often and more rigourously in an effort to avoid exposing thenselves to the perceived risks. This nervous grooming can be the cause of psychogenic dermatitis and alopecia.

I tend to see cases of dermatitis only after all medical potential causes have been eliminated and the possibility of psychological influence is all that remains. Often those cats are very similar in general disposition to James and, as with nervousness problems, treatment is usually a combination of modifying the cat's exposure to the stressful influence(s) as far as possible, while providing an opportunity to learn to cope under controlled circumstances. For many, a short period of hospitalization is a sufficient distraction to stop over-grooming and allow the essential treatment of the skin condition in a protected and non-stressful environment. In fact the slowing factor in treating psychogenic dermatitis is often the treatment of the dermatitis itself, which can be severe and, if it involves painful crusty ulcerations and secondary infections, can take ages to heal. In the process, the cat is responding to the itch by scratching and grooming the affected region, sometimes to the point where it has to be restrained from doing so by fitting an 'Elizabethan Collar'. Treatment of the original 'psychological' cause of the dermatitis is often assisted by a tapering dose prescription of sedatives or tranquilizers, similar to that used in treating more general nervous conditions.

The scratching behaviour may persist and maintain the condition long after the psychogenic cause has gone or the cat's competence has been improved. This can bring about a sort of self-reinforcing dermatitis which flares up when a series of stressful events arise. While we can alleviate intermittent bouts of stressful reactions using drugs short term, such cases are usually dealt with perfectly well at the vet's and are not persistent enough to warrant referral to me.

Self-mutilation

Self-mutilation is thankfully very rare in cats and in both cases I have seen, has been a continuation of psychogenic over-grooming in response to very severe stress. Having set up an eczema reaction on the skin, both cats continued to lick and scratch at the affected areas, past the ulcerated skin and down through their own flesh. Though thankfully still rare, it is more commonly seen as a reaction to stress in dogs than cats. Self inflicted damage to the flank area is an inherited disposition in many strains of Dobermann, particularly in America. Many Dobermanns elsewhere, my own dear 'Colonel' included, will suck their own flank in an uncomfortable position when resting or worried. Flank sucking will often cause eczema to develop as a result of the skin being frequently damp, but few cases go as far as self-mutilation down through the flesh. In cats or dogs, it is clear that the pain of self-mutilation is over-ridden by the activity itself as a reliever of the stress the animal is under, yet the sufferer remains just as sensitive to other painful stimuli. Powerful nervous blocks must therefore be released in the brain or central nervous system during self-mutilation which prevent pain avoidance responses. This in turn indicates how strong the animal's motivation to avoid the stress of solitude, enemies or other influences must be at the time it begins to mutilate itself. It appears that self-inflicted damage is preferable at the neurological level than merely the risk of damage from an outside influence. The continuation of the behaviour is probably connected with the brain's release of natural opiates as relievers of pain when the cat or dog is injured, masking the sensation of pain to such a level that the therapeutic angle of self-interested, self-directed behaviour continues.It is here at the micro-concentration level of neurological chemistry that the best hope for treatment lies. At present we can only try to remove stress and offer usual sedatives in treatment, although the use of the morphine antagonist Naloxone is being investigated in studies on horses in America and cats in Holland and may be researched at my Bristol University clinic in the future. Perhaps the first case will be one booked in for early 1990 of a Burmese cat which periodically tries to claw out its own tougue .

At my Bristol clinic I look forward to seeing cases of psychogenic dermatitis and self-mutilation with Mr David Shearer who has a special interest in skin conditions. Hopefully, between us we can get to grips with these problems using a combined veterinary/psychological approach to treatment. David gets the skin and I get the stress, but hopefully not the point of self-mutilation!

Psychogenic vomiting and regurgitation

Dear Mr Neville,

I have three wonderful moggies, two girls and a boy, all neutered. All are very different and have their likes and dislikes, but all are very affectionate with each other and me. The problem concerns the boy, Jasper, who vomits quite frequently when indoors. It seems to happen after the cats have had a small scrap or when I have visitors. My vet has tried feeding different diets, provided at different times of day, but to no avail. Could it be psychological? Is Jasper a compulsive vomiter, or is he just showing his dislike of some of my friends by being sick when they arrive?

Your sincerely

Frederick Reynolds

Many of us feel physically sick when frightened and when under stress of one sort or another. Fortunately few of us ever actually vomit except under severely stressful circumstances, and the same is true of cats. While most cats vomit occasionally or regurgitate hairballs with partially digested food, few are like Jasper. Such cats are usually rather incompetent generally and although not giving out other signals of being particularly nervous, live their lives under continual stress because they never habituate to what other cats regard as normal occurrences. As a result it takes relatively little to force them over the threshold of stressful reaction and a single rergurgitation of stomach contents at the time of peak stress, such as when visitors arrive in Jasper's house, or repeated retching and vomiting may occur as long as the stress persists and the cat is unable to avoid it.

Treatment involves steadily building up Jasper's confidence, as described in chapter 5 (pages 29 - 30) for the treatment of nervous cats, and trying to offer specific controlled exposure to those challenges that tip the balance. Such exposure ideally occurs when the cat has an empty stomach, so management of feeding times and the provision of light, easily digested meals that pass through the stomach quickly are also features of treatment. In Jasper's case it may just be that he can never settle in the company of other cats in his home territory and so the problem may be better managed by low dosage sedative prescription if rehoming is not considered an option by his owner. Where there is a prospect of such prescription being long term, I am becoming increasingly keen to replace the use of addictive sedatives such as valium with safer 'alternatives' such as homoeopathic or flower remedies under prescription or advice from veterinary surgeons with a special interest in these fields. The results with such cases so far have been very impressive, though the scientist in me still wants to know why. The important part of treating cats like Jasper is to produce some quick initial progress before Frederick runs out of friends who are less than impressed with Jasper's welcome, especially if they have come for dinner. The sound, sight and smell of cats being sick is perfectly designed to put you off eating and ought to be marketed as an aid to human dieting programmes!

Projectile vomiters

Dear Mr Neville,

I love cats and I am the proud owner of three, all of different breeds. Ming, my Siamese (neutered male), is always sick at night. But it isn't just a case of being sick , as when he throws up it is very liquid and there is lots of it. It comes out at very high speed and usually lands on the wall next to his bed. He doesn't seem to be in pain and my vet can find nothing wrong with him, but it happens nearly every night and either I have to get up and nurse him, or leave it running down the wall until morning. I can honestly say the last thing I ever need to see first thing in the morning is cat sick. Is this a psychological problem? My vet has tried his hardest with everything from antibiotics to a pyloro-myotomy (a loosening of the pyloric sphincter to ensure that food is not restricted in the stomach and can pass on normally to the rest of the gut) and suggested I write to you in case you had heard of it before and could offer any help.

Yours sincerely and rather tired of cleaning up,

Josephine Parker

No doubts about the diagnosis with this one, a projectile vomiter rather than a psychogenic vomiter, and a proficient one as well. It was difficult to know where to place this problem as it could equally well appear under the 'Bizarre' section later on, but since I treat it as a stress reaction due to lack of other sensible approaches from the behavioural point of view, here it is. Projectile vomiting can arise from a range of medical difficulties, but in this context is usually only reported in Siamese cats. Typically the problems start at around six months of age, then the vet exhausts himself for a year with drug treatments and surgery in trying to cure the cat and finally gives up. At which point the cat may head in my direction, usually at one to two years of age, and fighting fit!

Projectile vomiters are starting to get a strange hold over me as I find the behaviour irresistibly fascinating. How can it be that an apparently healthy, competent and sleek cat can have such a persistent and dramatic condition? Sometimes they are even slightly overweight, despite daily vomiting of at least some of their intake. Typically the vomiting occurs at night, four to five times per week and usually after a period of rest or sleep. While a few are single cats in the home, the majority come from two or three cat households. With one recent case, vomiting occurred so regularly that you could set your watch by it. Four o'clock in the morning. The cat was heard to retch for a few seconds with that amazing regurgitative noise that only cats can make when being sick, followed by a huge ejection of the partially digested and often very liquid food at high velocity. It usually lands on a wall with a splat. Not a pleasant sound and it certainly gives a whole new meaning to the words of the song 'Four in the morning'.

I now like to see such cases with Hilary Hill, a well respected veterinary surgeon who was the Feline Advisory Bureau scholar at Bristol when I started my clinics there and is still on the staff. Like me, she finds projectile vomiters extremely interesting, and we continue to liaise over treatment of vomiters past and present. Many projectile vomiters are succesfully treated by their own vet or by Hilary's expert attentions and surgery. She has found that some are caused by the overgrowth of certain types of bacteria at certain points in the digestive tract and can be treated successfully with appropriate antibiotics, but a few like Ming are unaltered by any such medical attention. I always refer these cases initially to Hilary for a full check-up in the light of her experiences. A few filter through to the cat shrink as a last resort. My success rate is 50 per cent in achieving a reduction in the frequency of the vomiting, but although this sounds reasonable, I should say I've only seen two cases, and we're not much nearer to understanding the true nature of this messy and worrying condition. However, both cases were highly memorable, and I'm keen to see more.

I have treated the condition as a response to stress from the presence of other house cats, even though there were no other signs of conflict and they were all happy to sleep together, groom each other, share food etc. Sometimes the presence of friends can be stressful and competitive even though they are liked or encouraged. The vomiting might be likened to the epimeletic vomiting of parent dogs or even certain seagulls which regurgitate stomach contents to feed their young. Although this behaviour is not reported between cats and kittens, it is about as close as I can guess to an explanation. Young puppies force their parents to vomit by worrying at their mouths until they get a meal and perhaps the stress and unobserved worrying by even friendly cats causes vomiting in cases such as Ming's. Why it should be projectile, caused by rapid constriction of the stomach, is more likely to be a genetic feature and restricted to the morphology of the Siamese. Overly complicated conjecture perhaps, but in terms of attempting treatment, very simple. We reduced all possible stress by isolating Ming and other sufferers during and after mealtimes, and offered a secure, warm room for sleep away from the others. Meals were given in small, digestable doses and only up to early evening to ensure that stomachs were empty by nightfall. In one case, sedative treatment was also prescribed for two weeks as an adjunct to help the cat cope with separation from its friends, as it was prone to be nervous when isolated at other times. With one of my cases, this has solved the problem even to the point where the cat can now be fed later in the afternoon and early evening. Other cases of Hilary's and the other with which I am involved continue to defy all treatment and we are currently exploring the prospect of keeping the cat more stimulated during the night by leaving it in a different room each night on a rota of five rooms with the windows open, the lights on and alarm clocks set to wake it up and stimulate it into moving to prevent prolonged relaxation. No results as yet; research is continuing into what I think is sometimes a very specific reaction to stress, but which is also the product of a combination of factors, physical and emotional, to which stress may contribute.

Trauma

Whereas stress is sometimes difficult to identify due to the large variation in individual responses to it and the apparent lack of any major or continuous behaviour change in many cases, trauma is manifested by marked behaviour responses. A physical shock from a wound is traumatic and causes the human victim to cry out, clutch the affected part and show other instantly fearful or defensive reactions to ensure that the damage is restricted and the prospect of repetition is minimised. Emotional trauma resulting from pain or even an unsuccessful attack can be just as dramatic, especially in cats, and can even lead to physical collapse and death, particularly with sensitive individuals of 'highly strung' breeds such as Siamese.

Interesting cases of 'shell shock' have been reported in cats which survived air raids during the last war. Similar to the reactions some people experienced after the blitz, the cats became withdrawn, unwilling to respond to their owners, and easily frightened by sudden movements or unusual or loud noise. As with other traumas one would expect that sound phobias developed afterwards in many of these cases, even where gentle handling and encouragement improved the cat's abilities to communicate and more normal levels of activity returned.

A more bizarre and hitherto unique behavioural response to trauma was witnessed in cats which were subjected to a flood at an English coastal resort. They subsequently made repeated efforts to paw at and pounce on imaginary objects on the ground and in the air, in a manner similar to the imaginary fly-catching behaviour of certain dogs. Cavalier King Charles Spaniels are particularly noted for this behaviour, but without the excuse of trauma. Unfortunately, no record of any treatment or the fate of the crazy coastal cats remains, so I must await my first case of this problem to see what might be done.

'Hello, is that Mr Neville?' a crackly voice asked on the 'phone one morning. 'I'm calling from the Middle East about my Siamese cats.'

I'd be fibbing if I said that I didn't immediately have visions of being flown to some fabulously rich, eastern potentate's palace, there to administer at least a month's advice on treating a forlorn hareem cat with a diamond collar. Alas, it was not to be. But the problem about to be related was of enormous interest.

'Mo is my mother cat; she's five and spayed and lives with Ali, her neutered son who is four. They've been together since Ali was born and we brought them out here with us two years ago from the UK.

They've always been totally devoted to each other; they sleep together, wash each other and even share the same feeding bowl. Then last week my husband accidentally stepped on Ali's tail. Ali yelled out and immediately ran to hide behind a chair in our living room, but was stopped by his mother who attacked him savagely. My husband got badly scratched and bitten in separating them. We kept them apart for an hour or so to let things cool down a bit and then reintroduced them. Bang! Off they went again with Mo attacking her own son just as aggressively as before. So we left them overnight and tried again but the same thing kept happening and has done every time we've tried to put them together again over the last week. Just the sight of him seems to set her off. Ali is obviously very frightened now, although both of them are perfectly normal with us on their own.'

We settled down to a long, expensive 'phone call to decide how to treat this very disturbing case of a trauma-induced breakdown in relations. It was ultimately resolved, but not quickly or easily, and not without hiccups along the way when progress in getting the cats back together was undone instantly by a flash back and repetition of the attack by Mo.

Trauma, by definition, comes as a surprise. The body's startle reactions to being attacked are swiftly by-passed. There is no opportunity for the flush of adrenaline to be put to use in helping the victim escape as the damage is done before the victim is aware of what is happening. Instead, adrenaline can cause the after-effects of shock; lowering of body temperature, increased heart rate pumping blood faster to constricted capilleries thus raising blood pressure, causing hair to erect and muscles to contract involuntarily producing nervous shaking. Dramatic yes, but thankfully rare as a cause of long-term behaviour problems in cats. They either die at the time from the physical trauma of a road accident, for example, or soon after from the effects of physiological shock.

Cases that I have seen of long term behavioural consequences of a single short traumatic incident usually concern relations between cats sharing a house, such as Ali and Mo, which can be totally broken down by something as fleeting as an accidental step on the tail by an unwitting owner. I am currently treating a similar case, also concerning two previously doting Siamese, whose relations were totally disrupted by a local rival entering the house via the cat flap. There was a single, short, serious fight with one of the resident cats. The other Siamese joined in, the visitor was swiftly defeated and chased out, but ever since this incident the first Siamese has attacked his friend on sight.

The reactions of the moment may be understandable in both cases. Ali, as a result of the shock and pain of being stepped on, cried out and perhaps this was spontaneously frightening to Mo. She saw him as the cause of her own anxiety and therefore instantly deserving of an attack to force eviction and bring relief to her own fear. What is surprising, however, is the permanent effect on their relationship. Simply the sight of Mo was initiating the same pronounced level of reaction in Ali weeks after the traumatic incident without any apparent diminution of her aggression with time or frequent exposure.

Treatment involved total control of their exposure to each other on a very gradual basis. Baskets were used to house each cat individually to allow them to share the same room for very short periods. They were placed at opposite ends of the large living room, with the owners present to offer what reassurance they could, they were allowed to react in their predictable way. Mo was frustrated in her attempts to attack Ali, who in turn was prevented from running away. Confined, he remained calm while his mother screamed at him and at least was unable to fuel her reactions by running. Once Mo's initial reactions had subsided the two were isolated from each other and the process repeated just two or three times per day for a week or so. They were left in the same room for progressively longer periods and more frequently as the weeks went on. On some days Mo would not react at all, and the cages could be brought closer together for a short period. On other days her reactions were extremely violent, even to the point where she had to be quelled with a water pistol, although there was a risk that this could have alarmed her more. Gradually, however, the trend was an improving one as mother and son learned to accept each other again.

They were released inbetween sessions in separate rooms and after about three weeks were allowed contact across a wire mesh divide between their respective rooms. Investigatory sniffing followed and things looked good, until . . . the husband went through the mesh 'door' and forgot to secure it behind him. One scream and a huge scrap later and we were back to square one. With husband suitably admonished and kept out of future treatment plans, we began again, and this time with steady progress over several weeks. When the great day came to put them together physically, it was actually done away from home in a friend's house so that both cats would be a little subdued in an unfamiliar environment. It seemed to work, and their mutual acceptance persisted when the process was repeated back at home.

The relationship between the two cats is still nothing like their original closeness, even many months later. They do not sleep together and are deliberately fed separately to avoid any escalation of a warning grumble over a favourite or remaining morsel. They are never left together unattended as there is little doubt that Mo would inflict serious and lasting damage on Ali if a fight went uninterrupted. She may even kill him, so we take no chances. Steadily, however, the two cats are getting closer again and from tolerating each other's presence will now wash each other for a few seconds before Ali, not sure of the wisdom of this, retreats slowly away.

There is no overcoming the effects of trauma by simply leaving such cats together to 'work it out', nor could any solution be achieved without much effort and patience on the part of the owners. The psychological effects of that one short accidental incident are far-reaching and probably permanent in many ways. Some owners would not have tried to repair the damage for too long and instead would have given one or other cat away or tried to ensure that the two cats never met again; perhaps this a sensible option in some cases. But at least it was interesting to see to what extent such a dramatic problem could be treated, and how long it can take even with such willing owners trying so hard.

8
Aggression

The reactions of Ali and others to trauma have often been markedly aggressive, with subsequent attacks directed towards a single cat simply as a result of catching sight of him or her. By contrast the traumatized coastal cats exhibited undirected predatory aggression responses to their flooding trauma by pawing at the air and chasing imaginary prey. While these cases are rare and rather specific responses to obvious cues, they indicate the breadth of the range of behaviour which we all too easily lump under a single heading of 'aggression.' When we start to consider 'normal' aggressive responses, that range becomes even broader. Perhaps we are too quick to label certain types of behaviour as aggressive because of the potentially dangerous consequences of attack.

We act defensively to avoid the aggression being directed towards us and this can prevent us from distinguishing between different types of agressive behaviour.

When dealing with cats, it would be extremely unwise for me to treat all cases of aggression similarly, even though the actual behavioural response of each cat may be the same, with the use of claws and teeth. Each must be treated individually, looking closely at what triggers the response and its scale, duration and direction, for treatment to have any hope at all. The general reaction of owners in the face of all types of aggression in animals is to become aggressive in return and try to subdue the cat or dog with greater force than being displayed, unless the animal is already beyond safe control, or is likely to inflict injury on them. In nearly all such cases, meeting aggression with excitement or aggression in return simply fuels the cat or dog's aggression and has little or no reforming effect.

More often, and particularly with canine aggression, owners are attacked or bitten when they needn't have been. While that may be their fault, the sad consequence is that the dog becomes aggressive more quickly when next challenged, or more readily diverts other forms of aggression onto its owner. The end result is an upward spiral of aggression with the relationship in tatters, owners bitten and dogs punished or even destroyed. Sadly the worst cases seem to involve owners who have received wrong advice from so-called dog trainers whose only answer to disobedience or defensive growls from dogs, frightened by over-assertive training techniques, is to clobber the poor dog with an ever larger stick. However, most trainers worth their salt recognize that there are many different forms of aggression. They know that to punish frightened dogs which growl in an effort to deter the perceived threat is bound to frighten them more and instead use calming techniques.

Enough of dogs and my concerns over the standards of dog training. My point was simply to show how important it is to differentiate between types of aggression and never to treat it by applying aggression in return. This is even more important with cats which, although far less likely to present problems of aggression than dogs because of that well evolved ability to avoid conflict, are even better armed with a paw full of claws at each corner as well as teeth at the front. Potentially they are far more likely to inflict injury on us than most dogs if we get caught up with an angry one.

But just what is aggression in behavioural terms? Aggression is carrying out an act of hostility or causing injury to a rival, or a prey item, or to a threatening challenge. Indeed the term 'aggression' is often used to describe only the first hostile act between two parties, though the victim may respond in like fashion and be described as defensive until he starts to win. Aggression is an integral part of every animal's behavioural repertoire, and it is expressed in a range of types and responses.

Body language

Predatory aggression is vital for the survival of all wild hunters, as are other forms of aggressive reaction for defence when challenged. But aggression in a social or interactive context between two adult cats or owner and cat is a different matter altogether. Indeed the repertoire of responses between two cats involved in a territory dispute may concern not only obvious displays of aggression in actual fights, but also a number of fearful responses designed to ameliorate the aggression in the competitor. The interplay between the two rivals is often enthralling to watch (providing actual fighting doesn't occur) because of the dramatic tension and the range of responses in both cats. Body movements and postures, facial expressions and vocalization are all used to communicate one cat's proposals or emotional state to another. The other will respond with equalizing reactions if standing firm, raise the stakes with more assertive gestures, or show retiring responses in deference to the other cat's assertive gestures. Such reactions are probably best described as agonistic rather than aggressive displays as they are actually designed to prevent violence. The bluff of staring your opponent in the eye, settling into a fixed tense position as if ready to strike, erection of hair, arching of the back and sideways presentation to make yourself look larger and more imposing, coupled with distracting tail swishes and low threatening growls or whines from a crouched position, are all statements of intent designed to convince the opposition to withdraw and so avoid actual physical interaction.

The squaring up of strutting and staring which often lasts only half a minute or so between agonistic dogs, can last half an hour or more between cats. Often two rivals will settle into prolonged staring matches from apparently comfortable positions, crouched on the ground, and vary the sequence and combination of their agonistic gestures and sounds until one edges away – almost, it seems, as much from boredom as discretion. The reason it takes so long for such interplay to prevent an aggressive attack from either cat, compared to the relatively short tolerance of high ranking or territorial dogs, is probably because of a lack of social cohesion and development of formal body language which keeps a dog pack together as a hunting unit.

Cats, being solo predators, have not evolved a social system of dominance/submission interaction for any purpose other than occasional territorial or resource defence. Instead most cats use that discretionary avoidance of conflict to ensure survival by keeping out of harm's way. Indeed, some authorities on cat behaviour do not recognize submission in the cat because it retreats when out-manoeuvred by a rival rather than rolling on it's back like a submissive dog. I think that gestures of submission are given off vocally, and through eye movement and body language prior to this, but we are as yet unable to translate the subtle combination or sequence of feline communications when they are all at work. Some evidence for this is that a slowly retreating cat is usually allowed to withdraw by the 'victor' without being attacked, yet cats which try to run away, a very submissive gesture practised by many losers in the animal kingdom, invariably stimulate the 'victor' to chase and attack to enforce the message. Or is it just that movement easily elicits aggression in cats and the roused, territorially agonistic cat simply switches into predatory chase aggression given a moving target? I doubt it, but as yet no-one seems to know for sure.

Aggression in terms of actual fighting between cats is rare because even a very confident dominant individual puts him or herself at risk from a defending cat's claws and teeth while attacking, so it is safer to resolve disputes with agonistic displays. Fights are more common between males, especially un-neutered ones. This form of aggression is predetermined by genetic influences as a result of the early masculinization of the foetus brain with testosterone hormone in the early definition of the kitten's sex. Later production of testosterone in the adult male cat causes the development of secondary sexual behaviour patterns, one of which is fighting with other males to establish a territory containing survival resources and breeding females. Castration removes much of the secondary hormonal motivation to this behaviour by removing the source of testosterone, as well causing the size of the cat's territory, and therefore the number of border disputes, to decrease. Most of us have our tom cats castrated at adolescence and don't own serious fighters (or smelly indoor sprayers, which is another largely testosterone-inspired pattern). Indeed I have never had an un-neutered tom referred to me for aggression problems, and those that I have seen are always stud pedigree cats kept safely in a pen and denied the opportunity to meet or fight other toms. Those owners who can live with an un-neutered free-ranging tom will have to accept that he comes home less often and may be frequently bloodied, bitten and scarred from fighting. Abscesses often develop from deeper bites and scratches, and owners must simply be prepared to patch their tough guy up for the next encounter. This always seems rough in high-cat-density areas in our towns, particularly as the cat gets older and less able to fight. Castration reliably reduces the motivation to be aggressive towards other males. Fortunately over eighty per cent of pet cats in the UK are neutered and so few toms are forced to suffer as a result of their hormones in the overcrowded territories of suburban catland.

Classifying aggression

The classification of aggression problems in cats falls neatly into two distinct types – 'normal', or expected under certain conditions, and 'abnormal', or excessive reactions. The concept of what constitutes any form of problem behaviour is largely a matter of opinion for each owner but, with aggressive cats, the opinion is more readily determined by the risk of injury and so tolerance levels are often lower. Perhaps for this reason I seem to encounter a greater number of cats presenting various forms of aggression than most cat owners would believe. Most cases do not concern normal or expected aggression, as most owners are aware of their cat's armoury and predatory status, but nonetheless a few cases do filter through of the aggressive behaviour that has been lost to most cats in the process of domestication and ordinarily is only seen in un-neutered toms.

As with many aggression problems it is not always the violence of the problem which causes the upset for the owners. More often it is the contrast with the cat's normally placid, friendly countenance that comes as such a shock, and it is the inability to intervene with the usual friendly affectionate overtures that causes owners to feel detached from their cat.

Predatory aggression

Dear Mr Neville,

My cat Arthur is an absolute murderer. He brings home at least one mouse, vole or bird every day and has even managed fully grown rabbits and pheasants. I know that cats are predators but is there anything I can do to reduce the level of his hunting apart from keeping him in all the time?

Yours sincerely

Maureen Beany

70

Predatory aggression is probably the most accepted form of aggression in our cats. None of us like it, most of us feel guilty about it, yet most of us realize there is precious little we can do to prevent it. Some cats are better hunters than others and a few are either not interested or just very poor at it. Some catch their prey, play with it, then lose interest and abandon it once dead – these are presumed to be non-hunters by their owners. Most, however, bring at least an occasional trophy home to their safe feeding lair, and some go as far as consuming most of their spoils. Hard rodent heads and a gall bladder left on the floor are all we usually find, though again, some give up once indoors and abandon their victim in favour of tastier, passive catfood in a bowl. When hunting, handling and despatching their prey, cats exhibit none of the prolonged staring, puffed body language or vocalization of the agonistic conflicts between cats or when fighting. Rather, it is a quiet, efficient affair with stealth and speed the key to success.

Recognized for their effectiveness in small rodent control since the settlements of the Ancient Egyptians, cats have been tolerated and even encouraged on farms around the world ever since. In 1868 in England, the Money Order Office in London requested two shillings per week from the Post Office to feed three cats under what became known as the Cat System. Only a shilling per week was authorized because it was felt that the cats should 'rely on mice for the remainder of their emoluments.' The Cat System spread to branch post offices throughout the country and it was still common to have a cat or two as official rodent controllers in Post Offices as recently as the 1960s. Many other companies and organizations have kept cats on the payroll for rodent control, and many have achieved fame for their prowess. In six years a female tabby was reported to have killed 12,480 rats at the old White City Stadium in London, and the place of the ship's cat is well established, even though the practice was discontinued in the Royal Navy in 1975 because of the risk of bringing rabies into Britain. Most famous perhaps are the Government cats. A black cat, always called Peter, has stalked the corridors of power in Whitehall since 1883. Winston Churchill's famous cat, Jock, kept him company and the rats and mice down in the underground war office, while the great man plotted Hitler's downfall, and the tradition of keeping a cat at 10 Downing Street continued until 1988 when the last incumbent called Wilberforce sadly passed away. At the time of writing, Mrs Thatcher has yet to replace him; presumably the post of

'Cat' in Downing Street is just another victim of government cutbacks.

Clearly cats have been officially appreciated in the very highest of circles for their hunting ability, though not all owners really like to face up to their pets' more bloody activities. Generally our attitude is to put such predatory behaviour down to the cat's basic instincts and to wash our hands of the carnage unless the cat is like Arthur and brings an enormous volume of carcases home. Others may bring them home alive and, having lost interest, release them to scurry or flutter around in the house, or even take what seems to us to be a sadistic pleasure in torturing a half dead victim on the lawn. Then we interrupt and, having grabbed the cat or chased him off, feel like St George as the small furred or feathered victim escapes to live another day.

An American biologist who decided to investigate the environmental impact of pet cats living in a small village in Bedfordshire came up with some alarming results. Over the course of one year, 70 free-ranging pet cats in the village of Felmersham brought home no fewer than 1090 prey items. Owners who dutifully complied with Professor Robert May's request to bag up each tiny victim for identification were informed at the end of the study that their darling cats had accounted for 535 mammals, 297 birds and 258 other unidentified small and furry organisms (UsFOs!). For the statistically minded this gives an average figure of 15.6 victims per cat, though this only includes prey actually brought home. The real figure including prey items killed and abandoned away from home or totally consumed will certainly be higher. One American study estimates that cats only bring half their victims home but if we extrapolate the figures as presented to cover the country's total feline population, we find that our pet cat population of 6,800,000 kill 95,288,400 mammals and birds annually, allowing for ten per cent of cats which live permanently indoors and have no opportunity to hunt. Similar data were produced by British scientists headed by Professor John Lawton of the University of London. Studying the predatory habits of seventy-seven cats in a village in Bedfordshire, they found over 1100 prey items were brought home, sixteen per cent of which comprised sparrows. They considered from this that predation by cats accounts for between a third and a half of sparrow deaths and that cats kill at least 20 million birds a year. This is interesting when we look at how outraged we are at the bird trapping and shooting by

European community fellow members around the Mediterranean.

The ecological impact of this pet predator is so startling because it fills a niche in our country and town ecosystems that is otherwise unoccupied. The only natural small predators we have are stoats, polecats and pine-martens, all far less common than even fifty years ago, although foxes and some owls are making successful comebacks after learning to adapt to the urbanization of the countryside. Our natural wild cat has been in serious decline for many years and is now probably lucky to catch 15.6 victims per day in its last areas in Scotland.

Bolstered by all that lovely food we give him, the pet cat is enjoying exercising his unforsaken predatory prowess to the full in the perfect hunting reserve of Great Britain, and as we acquire more cats as pets, we stack the odds against our small mammal and bird population. One last thought. In 1988 just over 2.7 million rats and mice were killed for experimental purposes. While we rightly feel outraged and campaign for the adoption of alternative testing techniques in laboratories and cruelty-free cosmetics, our cats are outside killing 35 times as many, and not even for essential food.

So, Maureen, I'm afraid that we are rather up against that uncompromising 13 million years of evolution since the first cat walked the earth in trying to reform Arthur's murderous habits. Even five thousand years of domestication, and selection for friendliness and beauty in the past hundred or so, has done little to alter his instincts. All we can try to do without frustrating all his outdoor activities by keeping him permanently indoors, which should never be regarded as an option with a cat used to the outdoor life, is to put him at as great a disadvantage as possible. A bell worn on his collar may give some targets a little more time to escape and restricting the cat's opportunities to hunt by keeping him in at times when rodent populations are most active, at dawn and dusk, may also reduce the carnage. Temporary boarding for a few weeks or so is not unheard of during the nesting season to protect young birds.

Another idea to calm the hunting motivations is to feed the cat a good full meal of favourite irresistible food before he is allowed out so as to slow him down a little. This I call the 'Alcatraz Tactic' after the famous island prison's policy of feeding inmates large volumes of fattening food to keep them unfit for escape and unlikely to be able to swim across San Francisco bay to freedom if they did.

Feeding the predatory cat fresh, gristly meat still attached to the bone forces him to spend much more time 'handling' and ingesting his meal, which may also reduce the desire to do it again with live prey. But perhaps Maureen will sleep more easily if she prevents herself from having to witness Arthur's 'crimes' by closing the cat flap and denying him direct access to the safety of his feeding lair. Couple this with shooing him away from the back door with a hiss and a small cup of water when he tries to bring a victim home and this may at least persuade him that the kitchen isn't such a safe place to bring his spoils. Better still perhaps is to obtain a kitten from a non-hunting mother and keep it indoors and unexposed to rodents and birds until about one year of age. This tactic is reported to produce cats with an undeveloped interest in birds at least and undeveloped skills in tackling rodents. However the lack of opportunity to develop in other respects may also produce a rather dull cat as a pet and if it catches up and becomes more interesting through new opportunities after it is allowed out, I suspect that in time it will learn to hunt, even if not as effectively. Movement of small beasts and hopping birds will still arouse chase instincts. I have treated several older cats (for other problems) which, when allowed out late in life following a move out of the city to the country, have quickly proved themselves to be very efficient hunters.

Dear Mr Neville

My cat is called 'Mouse', which is very appropriate because she is very good at hunting. I know it's natural but can you tell me why she has to play with the poor little things and torture them to death? Is there any way of training her to kill things more quickly and humanely if she must hunt?

Yours sincerely

Kay Brown

The usual hunting strategy of a cat is well known to us all. Attracted by the scuffling noise of a mouse in the undergrowth or the movement of a vole scurrying past or a bird hopping along the ground, the cat freezes momentarily and stealthily creeps towards it, crouching low to avoid detection. The eyes fix on the target, and with a quick aiming, muscle-activating wiggle of the rear end the cat achieves a lightning strike capture for about one in three small rodents and slightly fewer birds who have the advantage over the cat of being able to fly. The prey is grabbed with hooked claws and quickly bitten, but what follows is largely a function of the cat's experience. Mother cats which are proficient hunters raise proficient kittens, able to recognize prey items at about a month old after she has brought home a few items to smell, and then taste, in the process of weaning. She teaches her kittens the art of the swift nape bite to sever the spinal cord while keeping the rodent's teeth facing away from the cat and therefore unable to bite back. The nape bite is designed to despatch the victim quickly ready for eating. The ability is taught and developed through offering a wiggling tail to encourage chase and pounce by the kittens, and then the half dead prey is brought back to the nest for them to learn to handle and despatch. A few misplaced bites from the kitten earn a return bite from the victim so, through guided trial and error, he soon learns where to bite to effect the safest and quickest method of immobilizing the rodent.

Inexperienced mothers may not offer a wiggling tail very often and inexperienced or poor hunters may never bring enough half- dead prey on which the kittens can practise. The young cat therefore never progresses beyond an instinctive interest in movement. Such interest may even cause a cat to abandon a dead mouse to pursue a live one but, having caught it, interest may only be maintained as long as the mouse struggles and tries to run away. As the mouse grows weak from being pounced on repeatedly, the cat may even try to 'activate' it to maintain his interest by throwing it in the air, or allowing it to get away a little before chasing again with the apparent aim of encouraging the mouse at least to try to escape. Once the mouse is dead, however, many cats lose interest and do not continue to 'play' or go on to consume their victims.

In a world without free food in bowls such inept hunters would be ill-equipped to survive and only the proficient hunters would live long enough to produce kittens to which they can pass their skills. Perhaps in the last few decades we have been finally watering down the predatory side of the cat by breeding from mother cats with poorly developed hunting skills. But it will not be a short process. Many cats make up for early 'deprivation' and we deny most the opportunity to pass on their hunting incompetence to successive generations by neutering them.

Perhaps the best hope for producing a non-predatory cat lies with the pampered pedigrees, kept fertile for breeding in sheltered houses or pens without the opportunity to develop prey handling skills. But somehow I'm sure that even their kittens will retain the potential for interest in chasing mice, rats, voles, shrews, birds, rabbits and pheasants given the opportunity. As a result, I suspect that an even higher proportion would, like Mouse, 'torture' their prey never learning the swift despatch techniques of the nape bite, and so continue to disgust their owners given the chance.

Maternal aggression

Dear Mr Neville,

When Lucy my Birman queen produced a litter in May, she became very aggressive towards all of us and wouldn't let us near the kittens. While I understand that it is probably natural for her to defend her kittens, it seemed that she didn't recognize us and even attacked us when we were just walking by, not even trying to take a look. Friends who also breed cats said they'd never seen such a change in a normally docile, friendly cat nor such a defensive mother. Do you think think she was upset by this? Should we breed from her again?

Yours faithfully

Susan Swift

Maternal aggression is another of those acceptable forms of aggression with which we can readily identify and so tolerate, even though the scale of the attack on us in a different context may be totally unacceptable. The majority of female cats don't defend their kittens to such a degree as Lucy, but when they do it can be one of the most pronounced types of aggression we are likely to encounter. While most are happy to let us in to that most special of areas, the nest, others which feel challenged by our approach have to mute their own flight responses to stay to defend their genetic investment for the future. The pent-up emotion can then suddenly explode with a ferocious attack, without warning, either at us or other cats which are ordinarily tolerated or even accepted as friends.

Dr Benjamin Hart, from California, and one of the founders of animal behaviour therapy, suggests that such aggression may be released by the fall in the level of progesterone hormone at the time of birth. During pregnancy this hormone has been produced in large quantities to help the mother's body cope with the physiological changes involved. Progesterone also acts a calming influence on the areas controlling emotions in the brain and so with this capping effect lost the mother may bounce back to become highly reactive and easily provoked into emotional reactions and defensive behaviours. In humans the loss of progesterone levels is implicated in some mothers who suffer from post-natal depression so perhaps there is a link, and perhaps in cats some types of maternal aggression might be treated in the short term by the administration of artificial progesterone, known as progestins, from the veterinary surgeon. Otherwise it's a case of providing as secure a nesting area as possible for cats like Lucy and not attempting to handle kittens until some days after birth and only then when the mother is away, until she is more relaxed about our passing and approaches. Interestingly, however, cats like Lucy do not always respond in the same way with each litter and may be perfectly tolerant of people approaching her next kittens. Far from being over protective, other female cats with poor mothering ability may neglect or abandon their kittens entirely and then we are forced to take on the whole role of bringing them up. Again we are propping up incompetence in cats, for in the wild such poor mothers would never reproduce successfully.

Dear Mr Neville,

When one of my eight cats, Charlie, had a litter she ate them all after four days. Why should she do such a horrid thing and should we have her spayed now so that it can't happen again?

Yours faithfully

Pat Bream

Cannibalism is recorded in many species including mice, rabbits, dogs and, more rarely, in cats. Although abhorrent to us, it can have perfectly reasonable explanations. One may be a failure of the hormone system to inhibit killing of prey at the time of birth and shortly afterwards. Another, in feral colonies in particular, may be that the drive for self-preservation overcomes maternal instincts in malnourished mothers. Litters resulting from the second pregnancy in a year are apparently more prone to being cannibalized, as are those which are sickly and perhaps not worth the maternal investment to raise them, according to Ben Hart. Larger litters are more likely to be cannibalized but, more usually, cannibalism occurs as a result of the failure of the mother to find a secure nest area in which to nurse and raise her litter. Stress caused by overcrowding lies at the root of many cases of cannibalism and Charlie probably fits this bill. Next time she should be given a safe, quiet kittening area away from all other cats at least two weeks before giving birth and perhaps she will then take good care of her offspring.

Cannibalism of kittens by tom cats is often cited but is rarely experienced in pet cats. In prides of lions however, new dominant males taking over a group of females will systematically kill all cubs sired by the previous male in a ruthless purge. This was recently filmed for the first time in Africa and shown at peak viewing time to the British public to much debate about the acceptability of such violence at such a time, even if it was a true nature programme. The aim of the infanticide is to ensure that cubs with no genetic input from the new male do not drain his reserves in helping to fend for them and feed them. Instead the mothers of those cubs came into season again within twenty four hours of the massacre and were mated by the new male. They subsequently gave birth to his cubs, which, deserving of his protection, ensured that his success was passed on in the form of his genetic complement, not his rival's. Such behaviour is also reported to occur in the domestic cat when living in social feral groups, but rarely happens in our homes as in many cases they are protected from the roaming and perhaps long-gone father or, if resident, he is safe around his own kittens anyway. Nonetheless, a most important way of preventing cannibalism is to provide a secure resting area, safe from other home cats, especially males. This is crucial during the first few days after birth and can be relaxed with friendly cats later. Introductions should be supervised and if the mother is intolerant, permanent protection should be afforded until the kittens are weaned.

Dear Mr Neville,

When our cat Belle had her kittens we decided immediately that we would keep one particularly pretty daughter as company for her after all the others had been homed. Although Belle was an exemplary mother, she now won't have anything to do with Tina who is now six months old, and even spits at her and tries to chase her away if she comes into the same room. Why should Belle be like this to her own daughter? Is there anything we can do to make them like each other or should we look for a new home for Tina?

Yours sincerely

Caroline and Colin Barratt

Poor Caroline and Colin. It seems that Belle is one of those highly territorial females who cannot tolerate competition for food, shelter and perhaps most important of all, kittening area. While some females will accept any number of cats, related and unrelated, and even share in the nursing and upbringing of other kittens, a few are even intolerant of the presence of their own offspring once they are no longer kittens. There are none of the human family values of protecting children into adulthood with cats. While the kittens are dependent on their mothers for food prior to weaning and for protection and development for a few months afterwards, cats such as Belle make as good a mother as the next cat. But once their own offspring start to mature and become nutritionally and behaviourally independent, the mother will try to edge them out of her territory to establish their own and ensure she keeps all those resources that she will need to keep herself and any subsequent litters. While this perhaps indicates forward planning, the results are often spectacular with the levels of aggression demonstrated very marked. Occasionally a youngster may learn to fight back and establish a right to share in those resources again and even in time to enjoy quite a close relationship with its mother. However the chances of this are probably about equal to those of successfully introducing an entirely strange cat, and the cat will receive no special favours for being related to its mother.

In feral colonies the expunged daughters may establish territories next to that of their mother, and subsequent encounters at the borders may be reasonably amicable. My own observations of feral cats and experience of cases referred, would seem to indicate that male offspring are more likely to be tolerated for longer by territorial mothers than females, but there is no other evidence to support this. When female offspring become sexually competitive with their mothers the fur may fly, as then the resources for raising kittens are seriously under challenge. As males will not present this form of competition they may be tolerated longer, though ultimately they too are driven away as a natural mechanism to prevent inbreeding.

But for Belle and Tina the prospects of harmony are not good, and integration probably impossible unless both cats are spayed. But even then tolerance of each other may be all that can be achieved and the cats will never be seen to play or sleep together or groom each other. In fact, it is possible that they will never be relaxed when together and therefore the house will always be in a state of tension to the point where it is better from everyone's point of view to comply with the natural process and find Tina a nice new home. It's a pity, especially when so many cats do get on so well together and bond very closely and permanently to related individuals.

Food guarding

Dear Mr Neville,

Whenever I feed my tabby cat Campbell he crouches low over his bowl and growls fiercely. He eats in short quick gulps, pausing after picking up a lump of meat to look around. If I am close he growls again and if I try to stroke him he drops the food, hisses at me and has even lashed out at me with his paw. Of course I now leave him on his own to eat his meals, but is this common in cats? I had a dog once which used to guard bones but I managed to train him out of it. Could this be done with Campbell?

Yours sincerely

Simon Blake

Food guarding is more a feature of canine behaviour than feline. Bones are the ultimate tasty package and being none too easy to consume, invoke a guarding response to ensure that ownership is retained long enough to get at the goodies. Dogs are also more likely to guard high quality canned or fresh diets than less palatable dry or cereal based canned diets, and diet manipulation is a key feature of treatment. A few cats also guard their food, especially when in direct competition with other cats at the bowl. Ears may be flattened, growls emitted and agonistic body postures adopted but, by and large, the subordinate individual bows out without coming to blows. Standing low over the bowl and growling at the owner who has only an instant before gone to the trouble of opening the can and dishing it out to a hungry cat may seem ungrateful, but is a natural and easily explained agonistic reaction designed to preserve resources. Feeding more frequent small volumes of food which are more quickly consumed may lower the intensity of the reaction, as will providing continuously available dry diets, which involve not only less concentration of feeding opportunity but, being less aromatic and attractive, are less likely to be worth guarding anyway. As with dogs, however, all conflict at the food bowl must be avoided so Campbell and others like him are best fed on their own and left undisturbed.

Pain-induced aggression

Dear Mr Neville,

I have tried to groom Snowflake, my Persian cat, every day since I bought him at 13 weeks of age. Now he is a year old and still hates being groomed. He wiggles and curses and then scratches and bites me. I've tried to be gentle and offer him titbits but nothing seems to calm him. After a while he is so wound up that he grabs my hand and arm and really bites hard. If I hold on, he bites harder, if I let go he runs away and won't let me near him. Is there an easier way? I really do have to groom him otherwise his fur gets tangled and matted.

Yours scratched

Helen Ford

Difficulties with grooming or trying to treat an injured cat are the main occasions when we are likely to encounter pain-induced aggression. It is understandable that injured cats will try to avoid further threat and make us keep our distance. Pain elicits an aggressive reaction as we know if we accidentally step on the cat's tail or inadvisedly smack a cat for a misdemeanour. Often as not we get an immediate swipe in return, and our efforts to correct the cat's behaviour fail because punishment has no therapeutic role with the adult cat. If the shock is too great, the cat may also be traumatized for some time and avoid our efforts to make up through a prolonged fearful reaction. Traumatic painful episodes can lead to low threshold, specifically targeted aggressive reactions as discussed in the Trauma section of the previous chapter (pages 61 - 65). Prolonged exposure to a less startling painful stimulus can also produce a progressively decreasing reaction as the cat habituates to the pain, and realizes perhaps that any counter aggression to it brings no relief. Early play fights between kittens help them control their responses to pain and associate their own biting of a litter mate with a painful counter-reaction of aggression in return. Kittens brought up in solitude do not acquire this awareness of the influence of others and consequently are often spiteful as pets and easily roused to aggressive attack when handled by us.

Grooming, however, is a different matter, especially when we've tried to be gentle in what, with long-haired cats, is an essential daily aspect of care. While some enjoy the attention and handling, others like Snowflake never accept it readily, and like the solo or hand-reared kitten are always aggressive in defence. The cause almost certainly lies in the cat's upbringing. Snowflake's breeders probably didn't acclimatize him properly to being groomed from the first day they could handle him, when those feedback mechanisms were open for manipulation. The fractious young kitten can be disciplined with a firm 'no' or mild counter tap if he lashes out, in the same way as his play aggression releases a corrective, aggressive response from his littermate. Tolerance develops and this can be rewarded with gentle handling and later on, with titbits and treats. If this process is delayed, the opportunity is missed and for evermore the cat will perceive the brush and comb as a major threat, always deserving of an aggressive response. As a result many owners have to resort to sedating the cat for the purpose of grooming, or worse, leave it as long as possible to get dirty and matted, and then have the cat clipped or shaved under anaesthetic by the vet.

Though the days of getting Snowflake to enjoy being groomed may be long passed, there are nonetheless alternatives to sedation in making grooming safer for the owner. First the cat should be placed on a perfectly steady table at a comfortable height for the owner. On the table should be placed a covering of thick carpet. The cat should be pulled gently backwards as instinctively he will try to hang on by extending his claws into the carpet. It may be a two person job, but the initial grooming should then be very light along the back and not directed at sensitive areas such as the abdomen until much later. Matts should also be avoided in the first few 'mock' grooming sessions. Gradually the brushing can become more penetrative, always accompanied by gentle voices and the prospect of treats, though it has to be said that few cats will be very interested in them until more tolerant of the handling. As with cases of aggression in kittens and young cats, where owners simply must persist with their approach to grooming or handling an adult, a steady increase in frequency and intensity of handling is most likely to subdue the cat's aggressive responses compared with occasional sudden, high pressure efforts. But when this tactic fails or is still providing the owner with scratched arms, a hood over the cat's head will have a calming influence and at least enable the job to be done, even if the cat is not actually learning to accept the handling. Recently launched on the pet market is the Kalmcat muzzle from Mikki, a specialist animal handling equipment company based in Luton, Bedfordshire. The Kalmcat comes in two sizes to fit either slim-faced orientals or broader faced moggies, and, while it does not hold the mouth closed, it does neatly fit over the cat's face and hood him. The typical reaction of the hooded cat is to crouch low and only move very slowly, if at all. Gentle grooming can then be safely introduced, as can other forms of invasive handling such as the administration of eardrops, to which cats often respond with aggression if able to see. Similar responses to darkness are also utilized by hooding birds of prey, and to calm horses which are reluctant to enter the stalls at the racecourse. Ideally, however, all kittens should be handled frequently and, especially if long haired, groomed as early as possible and well before six weeks old.

Stroking and aggression

The strange nature of the threshold between tolerance or even enjoyment of handling, and aggression in cats is well illustrated by what the Americans, in their urge to make a slogan out of everything, call the 'Petting and Biting Syndrome' or 'Petting Induced Syndrome'. Most affectionate pet cats can be encouraged to exhibit this type of aggression quite readily.

The cat, which in America at least is more likely to be male than female, apparently enjoys a cuddle, settling comfortably in its owner's arms and purring loudly. Stroking the head and back could go on for ever, but touch the abdomen or tickle the hind legs and nearly all cats will suddenly grasp your hand or arm, scratch and bite, and sometimes kick with a repeated movement of the back feet. Others respond in a similar fashion to far less physical interference, sometimes for simply trying to pick them up. Most cats will at least tolerate being picked up and enjoy being cuddled for a short period but, from lying in your arms relaxed and happy one minute, the cat may become instantly violent, looking almost as confused as us. As we curse or yell at the traitorous little so and so, he has already jumped down from our arms or lap and may stand a few feet away, with ears back and not sure of his next move. Often it will be to sit and groom enthusiastically, which is sometimes a diversionary self-directed behaviour that cats enact when confused or slightly nervous and sometimes an effort to wash or spread the smell he has collected from your hands more evenly over his body.

For cats which are used to our nursing attentions the threshold of this reaction may be very high, and only rarely encountered. For others, perhaps unused to handling or handled less frequently as kittens, the reaction appears within a few moments of being petted and without us having to touch those more sensitive areas. It seems that the threshold of the reaction is reached when the cat ceases to feel comforted by our mothering-style affection and suddenly feels trapped and vulnerable so close to us while in such a relaxed state. From having allowed himself to revert to his kittenhood in that maternal/kitten relationship we encourage, he grows up in an instant and as an adult solo predator, decides that he needs to repel what has become a mild threat and make a distance between us. Small wonder he looks a bit confused as he jumps down.

Theoretically the potential for this reaction lies in every pet cat. Treatment simply involves trying to raise the threshold of the reaction by never reaching it. Cats with low thresholds of reaction should be fussed in very frequent, very short encounters, initially not even involving picking them up. If they will accept this then they should be encouraged to lie sphinx-like on one's lap and be stroked on the head and along the back. Escape is easier from this position and does not involve having to beat us away before the cat can get down. Gradually the length of such contact can be increased so that the cat learns to feel more comfortable in physical contact with us, but only much later should attempts be made to invert him to be cuddled. Those sensitive areas should never be touched and the cat never restrained if he decides to depart, as both of these actions will lower the threshold of 'Petting and Biting Syndrome'.

Learned aggression

Dear Mr Neville,

My little female cat Rose is usually very quiet and affection-ate, but whenever she sees my neighbour's dog, or any other dog for that matter, she immediately attacks it. I thought cats were traditionally afraid of dogs, but where I live it's the other way round! Why should Rose be like this, especially to the dog next door who really only wants to be nice to everyone?

Yours faithfully

Sarah Plumb

Dogs and cats! From Tom's encounters with Spike in *Tom and Jerry* to the Peke-Faced Persian in the USA, which is a bizarre attempt on our part to select a cat which resembles a dog, to the not infrequent, but certainly impossible claims that someone has managed to interbreed the two species, we wonder about the relations between our two favourite pets. In cartoon town, 'Dogs hate cats' as Spike says when instructing his pup in one of those marvellous sketches, yet many of us who keep both know that they can be the best of friends. When problems arise it is usually from the canine end, as many dogs will chase cats or anything else that looks like it might run.
    Faced with a chasing dog most cats try to avoid the conflict and

run for cover as fast as they can, preferably over and behind a fence or up a tree where, once safe, they glower, highly displeased. A few learn that running away only heightens the dog's predatory excitement and stand their ground with ears back, body arched, threatening to strike. There is nothing quite so funny, as the makers of all those cartoons are aware, as when the charging dog suddenly realizes that this is no ordinary cowardly cat and that the chase should stop. The problem is stopping in time to avoid upsetting the cat and the roles are instantly reversed. The dog, his pride in tatters and glancing around to ensure that no-one witnessed his failure, withdraws and trots back to his owner while the cat settles to whatever it was he was doing before the dog appeared.

Unless, that is, the cat is like Rose. Perhaps a dog did just nip her one day, or perhaps she was sufficiently frightened by a swiftly advancing menacing mutt, or cornered and forced to beat off the challenge by unleashing a defensive swipe with a fully clawed left hook. The success of her aggressive defence saved the day and became incorporated into her behavioural repertoire as a learned pattern of aggression. But now, rather than waiting for the threat to manifest from an apparently relaxed non-threatening dog, she wades in first to force a retreat before he can attack; a pre-emptive strike to preclude any possibility of being caught at a disadvantage. Now the boot is on the other paw, and it is the dog which must learn the techniques of conflict avoidance. If necessary, treatment may be possible in acclimatizing Rose and her ilk to regularly encountered individual dogs such as the one next door, by penning her close by and allowing the dog to approach and sniff without being chased off. This is similar treatment to that of desensitizing cats to visitors described in Chapter 5. But while this may be successful with the one dog, it may not be so for all the others who dare come within range of cats like Rose.

Dog chasing is just one example of learned aggression in cats. Another is the ever quickening response of aggression to having your tail pulled repeatedly. This can be rather dangerous with young children about. Although the cat usually learns to find high sleeping places after a couple of assaults from a crawling or toddling child, the child may still pursue the cat as a moving item of interest every time he appears in the room. It is natural for the child to show that kind of interest, but it can be a frightening phase for the cat. The child wants to make contact but has little control over the strength of its patting or grabbing. The cat usually learns quickly that an aggressive swipe or bite deters such unwanted attention as the child screams in pain and stops advancing.

As with the chasing dog, the cat may soon learn to get in first and advance on the child to preclude all possibility of heavy handling. The cat perceives this as being a better option than trying to avoid the conflict in the usual way by running off. In some cases this type of learned aggression can be very hazardous indeed for a while, but may gradually dissipate as the child grows up and gains more control of its movements. The cat often comes to accept normal petting from the child later in the normal way, but a few remain aggressive at the sight of children, and not just those in the cat's own family. Visiting children may be under just as much threat and so in these cases the cat will need to be separated from family life to prevent injury.

Of course no cat or dog should ever be left alone unattended with a child for both their sakes but while such cats can be managed at home with a little care, it is probably safer in most instances to rehome the cat to a child-free house. Physical punishment of the cat by the child's parents only serves to heighten and spread the cat's aggressive reaction and will have no reformative or reconditioning benefits at all. Intervention from a dominant member of the group to induce a submissive response does not work with cats in the same way as a well timed vocal or startling intervention may do with a dog in a similar aggressive frame of mind. Aggression from us causes the cat to be fearful and lash out to protect itself even more.

Dear Mr Neville,

When we first got our cat, she was only a little scrap from the local animal shelter but soon fitted in well with our house and went on to carve out a territory in the local gardens. She bonded extremely closely to me and initially accepted visitors and my boyfriend very well. She was always quite rough when playing and seemed to be more ready to scratch and bite my boyfriend than anyone else. Since then I've changed my boyfriend and he has moved in. He adores cats but Chloe hates him despite every effort on his part to be nice. She physically attacks him whenever he walks in and if I hold her, she glares at him and growls. Now she doesn't like visitors either and attacks people she used to like before, even though she is still very affectionate to me, to the point of following me around the house and trying to sleep by my head at night. My problem is how to reconcile Chloe with my boyfriend and quickly because we plan to marry in a few months time!

Yours sincerely

Christine Allan

We have dealt with such specific dislikes earlier, with cats whose reactions are nervous and withdrawn. Specific aggressive reactions are less common, though every cat will have its likes and dislikes of different family members and visitors. To some extent Chloe's aggressive threshold may be low because of encouraged rough play when young, and success of the ultimate kind in having disposed (perhaps in her mind) of the previous boyfriend as well as learned success at repelling visitors with an aggressive display. The aims of treatment are essentially to acclimatize Chloe to the boyfriend gradually, perhaps with her protected initially in a pen or on a harness so that she can grow used to his right of access and presence. Christine, too, has a role to play in detaching herself from Chloe and relinquishing the roles of feeder and affection-provider when her boyfriend is with the cat. Initially frequent short meals should be the only advances made by boyfriend to cat and he should not attempt close handling until much later. Low and decreasing doses of tranquilizers of the type more commonly used to control car sickness and bonfire night anxiety in dogs seem particularly helpful in such cases, provided the cat receives frequent short exposure to the boyfriend. So if marriage is in the offing, this should be an aid to progress and the boyfriend need not be traded in for one that Chloe finds more acceptable!

Idiopathic aggression

Dear Mr Neville,

Our Burmese cat, Gong, is a pleasant friendly cat for most of the time. As we live in an eighth floor apartment he never goes outside but he enjoys a good life with us and receives much attention and petting. We give him only the best food and lots of toys to play with but approximately once a week he attacks either my husband or myself as we walk into the room where he happens to be lying or sitting. The attacks are quite fierce and aimed at our legs or arms. Gong seems manic during the attacks, which last only a few moments, and he takes some time to calm down afterwards. We live in fear of him sometimes but at other times he is the most responsive cat there could be. Is there a way of making him safe?

Yours sincerely

Clare and Steven Hall

Nearly all the cases of ostensibly motiveless aggression that I have treated concern Burmese cats that live permanently indoors. The majority of the others are moggies but, again, are indoor cats. While it is unwise to rule out the possibilities of fear-induced aggression in such cases, perhaps caused by the owners' entry into the cat's sleeping area and startling him, most do not fall into this category. The aggression is what scientists call 'idiopathic' which means 'no known cause'. In short we haven't got much of a clue why it happens, but we couch it in impressive sounding language to cover up.

There are medical conditions which may cause a cat to react in this way, such as hypothyroidism, and in this case thyroid hormone replacement therapy can resolve the problem, but this is obviously a job for the veterinary practitioner. Other causes include forms of epilepsy which can be successfully treated with anti-convulsants, and – rarely identified, but probably more common than we realize – hypersensitivity to certain foods and allergy to preservatives and anti-oxidants therein undoubtedly exacerbates, if not causes, some problems. We wait for science to catch up with our observations and explain more of the effects of diet on human and animal behaviour, and in the meantime try to eliminate this possibility as a cause of aggressive behaviour in cats and dogs by feeding consistent, balanced and preferably fresh diets, though the higher quality dry diets can be more helpful with canine aggression cases. Canned foods are out immediately, however.

Ostensibly incurable, motiveless aggression in cats may in some cases turn out, on post mortem examination, to have been caused by certain types of brain lesion, perhaps a tumour, or from scar tissue resulting from an infection. This is especially suspected if the lesions are discovered along the midline of the brain and in certain areas of the hypothalamus, where electrical stimulation of live cats invokes an aggressive response. However this does not help the live pet cat and its owners very much, at least not until brain surgery of cats becomes a viable approach to treatment.

Treatment of cats like Gong can, however, be very successful even if the cause of the aggression cannot be accurately determined. According to some animal ethologists or behaviourists, aggression is a kind of genetically predetermined life force, the periodic expression of which is almost obligatory from time to time to keep the animal rational at other times. Other theories suggest that it is only the capacity to be aggressive that is genetically pre-programmed and that certain stimuli are required for actual aggression to be released and for the most part its direction will be towards that stimulus. In the light of either theory it is logical to suggest that a highly intelligent self-determining creature such as a cat requires a certain amount of stimulation, both mental and physical for the release of energy and emotions. Never is this more true than with the Burmese breed which is renowned for its intelligence and demanding nature. Kept inside in a consistently unchallenging environment, however comfortable and well fed, it is not surprising that it will take very little to excite the cat. Against a consistent backdrop of soft unstimulating comfort the cat will be all over the owners when they come home because their arrival is the major change of his day. All his energies will be directed at them at this time and any other time when they care to make a fuss of him. But Gong will also tend to over-react to other stimuli which most cats would take no notice of. Clare and Tony's entry into the room is such a massive change to the norm, that Gong over-reacts in a major manner with hardly any assessment of the situation and certainly no recognition of his owners until he has inflicted serious injury. Once the energy has been released a little, the cat will often perceive himself to be at risk with his teeth locked onto his owner's leg, and will run away just as quickly. He may look confused afterwards and engage in some of that unnecessary, but tension-relieving, intensive grooming.

Treatment is quite simple. We have to provide much more stimulation in the cat's life when the owners are not there so that this situation does not arise. Ideally the cat should be allowed to go outside and once I have assured the owners that they are possibly being cruel by keeping the cat indoors, most who can, agree, and are never attacked again. The cat gets all the stimulation it needs through hunting, reacting with its environment and other cats, and returns home for food and affection.

This is not always possible for apartment dwellers like Gong, though I have known of owners who built very smart access ramps zig-zagging down the outside wall for two floors to allow their cat access to the garden. Other owners do not wish to expose their cat to the dangers of the outdoors and in areas with busy roads this is entirely reasonable. For these cats we must provide plenty of other forms of stimulation and this comes most readily in the form of another cat. Properly introduced, the company and sheer presence of the other cat can relieve all those emotions in the aggressor and make life safe for everyone. Interestingly I employ a similar technique when treating feather- plucking parrots. This is a stress reaction to boredom, and in severe cases can lead to what is known in parrot circles as the 'Buxted' oven-ready look. Providing another parrot in a cage alongside the sufferer is often all that is needed to alleviate the boredom and cause the feather-plucking to cease. In most parrot cases the two birds hate each other, but simply having competition to scream at and keep a beady eye on is enough to keep the feather plucker busy. With aggressive cats we hope for, and usually get, at least a tolerant relationship between two cats, with occasional play and the odd disagreement all being part of the enriched scene.

For owners who can't or won't take on a second cat, I suggest that they get their cat accustomed to wearing a harness and be taken out for walks like a dog at least twice a day. The cat should also be offered as many novel objects as can be found every day. They need not be expensive; simple cardboard boxes, branches, toys etc, all keep his curiosity exercised, and the owners are also advised to play intensively with the cat for an hour a day by making him chase moving toys in simulated hunting activity. But in many of the cases of solitary, confined and rather unstimulated cats, their general activity level can be managed from time to time with low doses of valium or progestin treatment. Even better is the use of alternative treatments like Bach's Flower Remedies. But always at the back of my mind is the worry that certain cats need to be offered the outdoor life to preclude the emotional build that leads to outbursts of aggression and while we can substitute, stimulate and moderate, what is really needed is for the cat to be rehomed to a more fulfilling lifestyle, and the owners to try again with two individuals of a less assertive breed perhaps better suited to life as apartment cats.

Redirected aggression

Dear Mr Neville,

Tyson is my big bruising tabby cat. He is a neutered boy but nonetheless is always fighting. Thankfully he usually wins and chases every other cat away from the garden. But sometimes he will sit by the patio windows and simply watch over his territory. If another cat walks through the garden or even along the fence (why they still do it I don't know) Tyson becomes extremely agitated. He growls like a dog or mews, his mouth trembles, his hair sticks up and his tail swishes angrily from side to side. I can understand why, as he hates all other cats being on his patch, but when I try to comfort him he attacks me immediately. Last time he bit my leg so badly that I had to stay in hospital for over a week. But I love him dearly despite this because he is usually very affectionate with me, when he recognizes who I am! What can I do about his aggression towards me?

Yours recovering

Marina Sargeant

If it's not going to be a Burmese, it'll be a tabby! While there is no real data on the disposition of any colour types towards certain behaviours, it does seem that while most tabbies are gentle and affectionate, despots like Tyson do occur, and if there is going to be a single dominant tyrant in a feral cat colony, in my experience at least, it is nearly always a tabby. Old English Tortoiseshell colours (that very attractive ginger and black marbling found only in female cats) are renowned by many cattery owners for being spitty and not easy to handle while many claim that black or black and white cats tend to have friendlier natures than the average cat.

Tyson's behaviour is a classic example of redirected aggression, and in this case, a very dangerous example. The excitement build-up at the sight of a rival daring to parade in front of Tyson has no release because he is frustrated in the desire to chase the other cat away. Tyson is fixated and studies every move with such intensity that he does not, as would be expected, immediately go the back door to ask to be let out, as he would when needing to relieve himself. Instead he sits and fumes. His owner, seeing his frustration, tries to comfort him, but unwittingly her approach triggers the release of his aggression, perhaps because he hasn't noticed her behind him and is momentarily startled by her, perhaps because her movements alone tip his aggression over that threshold of control. The severity of his attack is unusual and while the immediate injuries from teeth and claws will be severe, the stay in hospital was required because of likelihood of cat bites and deep scratches becoming infected. Teeth and claws are coated with all sorts of unpleasant bacteria, but particularly the *Pasteurella* species which when injected deep into our tissues can cause septicaemia. A long course of antibiotics and rest is required to fight off what can be a very nasty reaction. This injection of bacteria is also the reason why so many injuries to cats from fights with other cats turn into those purulent abscesses. But if it's any compensation, in the UK at least, there is only one mammalian bite loaded with more potential after-effects – not a dog's, not a rat's, but that of a human.

As for Tyson, greater access to the outdoors and encouragement to spend more time out there by installing a cat flap will help ensure that his aggression is directed towards his rivals. Keeping the curtains drawn over the patio windows and other lookouts to block his line of vision may help, and offering a warm attractive bed well away from windows may direct him away when indoors. Most important is that Marina does not seek to reconcile him physically when agitated. Instead she should employ distraction techniques from a distance by rolling a ball of paper or foil past him to encourage chase behaviour, by rattling the treat box or by opening the back door and calling him to develop that desire to go out and protect his territory at such times. Marina shouldn't walk too close and if distraction is impossible she may be wiser to sneak quietly out of the room by the nearest unblocked exit. Low dose prescription of tranquillizers from Tyson's vet may keep him less reactive but, in the long term, alternative remedies may be preferable. It will also be worth testing the effects of diet on Tyson's reaction threshold by weaning him off premium quality canned food for a week or two and offering fresh chicken or fish instead. However, this is really to extrapolate the observed effects of high protein diet sensitivity in some aggressive dogs. It may make little difference to cats.

Territorial (dominance) aggression

Dear Mr Neville,

My boyfriend Martin has two Siamese cats which live permanently indoors with him in his apartment. One cat, Suki, is placid and adorable, the other, Minkey, I have never really trusted. He has always eyed me rather carefully, as if waiting his chance to attack and I have always felt uneasy on my own with him. He got his chance this weekend. I got up in the night, and walked naked out of the bedroom to use the bathroom. He was sitting in the narrow passageway and in the half light I stopped to wonder whether I should walk past him. I decided to proceed thinking that it was only dogs who guard places. I was wrong. I had taken two steps forward when Minkey sprang at me, leaping on to my bosom and biting and scratching me severely. I screamed in terror and he ran away. Sometime later my boyfriend appeared but refuses to believe that I did no more than walk towards the cat. I had to go to the casualty department of the local hospital and have been on antibiotics ever since. My left bosom is badly bruised and has two distinct cat mouth-shaped bite marks on it. Needless to say I am refusing to go to his apartment again until Minkey is reformed, rehomed, or preferably destroyed, and setting a date for the wedding is unthinkable.

Yours sincerely

Carol Summer

Such directed aggression towards people by cats is rare indeed; in fact, this is the only case I have ever been called on to treat. Other aggressive cats of this nature are probably not given the opportunity to repeat their performance, and it is true that most men in Martin's position would be faced with the 'it's me or the cat' ultimatum and choose the former without delay. However, Martin was a little less keen to go through with the mechanics of parting with Minkey and he thought that, in time, the cat would accept Carol's arrival. He may have been right. However following the attack, Carol naturally became ailurophobic (ailurophobia or aelurophobia is a fear of cats; the word arises from *ailuros*, ancient Greek for ferret, as both ferrets and cats were kept as rodent controllers and somehow the words became mixed up). So it was no longer just a case of trying to get Carol friendly with and accepted by Minkey, but also a case of overcoming her fear or reluctance to be near any cats at all.

Ordinarily I would have advised that Minkey be rehomed, and that Carol try to make headway with Suki. But, in this case, it seemed that Martin was not prepared to countenance such a loss of bachelor ties with a faithful friend and so Carol decided to break off the whole affair. Then we had a lucky break. Martin moved out of his London apartment to the Hertfordshire countryside and the prospect of allowing Minkey free range outdoors became possible. In his new home Minkey became far less tense and far less attached to Martin, treating him more normally as a provider of food and as an occasional lap to sit on, rather than the centre of his existence. Monopolization of key areas of the house, and in particular narrow stairways, became far less important to Minkey. In his old apartment, Minkey had been behaving in a very dog-like fashion. More than that, he had been behaving like a very dominant dog, and at certain times expected the right of control over pack mates as to where and when they moved around. I treat many such dogs which guard the passage ways of stairs, hall corridors, doorways and, at night the entrance to the loo. With no room to walk past and maintain a respectful distance, innocent people approaching and trying to pass must come within the animal's personal space. While dogs usually growl to force a retreat, Minkey simply leapt up at Carol as soon as she made it clear that she was continuing her advance towards him. The consequences were extremely painful and unpleasant, and occurred as a direct result of the confrontation over the right of passage in a narrow enclosed area. Unusual, even unique, but not worth the risk unless the cat's desire to dominate and maintain rank could be dissipated.

This could only really be achieved if home could become a sleeping and feeding lair for Minkey, and this was the result of the move to the country. Alas it did little to overcome Carol's fears, but with new agreements regarding Minkey not being allowed in the bedroom, being shut in the kitchen on his own bed at night, not being left in the same room as Carol alone and the use of a protecting indoor pen for him to allow safe close contact with her, the couple got back together.

Social conflict

Dear Mr Neville,

Recently we bought a little kitten and brought her home to meet our two-year-old neutered, very affectionate male cat Bertie. We tried to fuss him a lot during their first meetings but even after many weeks of trying, Bertie will not accept Suzy. We have to hold him back to prevent him attacking her and he has even scratched us in his efforts. On the couple of occasions that he has got to her he has bitten her neck and grabbed her with his paws while kicking her quite severely with his back feet. On one occasion, having grabbed her neck he began to mount her which would seem to be the opposite reaction. Is Bertie confused and is there any hope of getting him to accept her?

Yours sincerely

Jenny and Julian Wright

Socializing a new kitten or cat with existing cats in a house is one of the most common problems that I am contacted about. It is also one of those problems where it is very difficult to predict whether treatment will be completely successful or not because every cat is different regarding what other, if any, it will allow to share its home and to what level. There is always the presumption that all cats can and should get on. After all, many people have nine, ten, eleven and more and they all get on quite happily and readily accept newcomers. Yet, like Gong of the misdirected aggression case, there are always some who simply will not tolerate another cat in their outside territory, let alone indoors. It all comes down to the key to the cat's success as a species.

Solitary as a hunter and unable to co-operate with others to co-ordinate a hunting strategy, yet retaining the ability to share resources and avoid social conflict when everything is plentiful. So why, when we provide everything in plenty, from love and protection to bowlfuls of food, shouldn't a cat let others in?

As we might imagine, there are several important influences to consider. The cat which has been kept with lots of others since it was a less competitive kitten and has been used to meeting and sharing resources with others, is more likely to accept a new arrival and tolerate it well after an initial introduction period. Full acceptance and friendship may take some time while the newcomer acquires the communal smell of the resident cats, dogs and owners. The resident cat must also have time to weigh up how the owners and other animals treat the new arrival; their acceptance and relaxed behaviour may help his own. In many respects kittens are more likely to be accepted than adult cats because they are not sexually or particularly socially competitive. A bite on the tail from a kitten is like a kick in the shins from a three-year-old child; although unpleasant and worthy of rebuke, it is not nearly as challenging as exactly the same behaviour from an adult. It will also help to reduce the competitive profile of young newly introduced cats by having them castrated or spayed as soon as advised by the vet. I well remember the case of a marvellous family that I saw at my clinic at Rutland House Veterinary Hospital in St Helens. They owned a beautiful young ginger tom which, after being accepted by the original black and white three-year-old neutered male cat, had suddenly and seriously been attacked by him. On being asked for details of the attack, the eight-year-old son replied, 'Well, Blackie keeps on marmalizing 'im'! While it seemed appropriate in some ways that this process should happen to a marmalade cat, the advice was straightforward: have him neutered without delay. Sure enough relations were restored between the cats in a matter of a couple of weeks. It was simply that Ginger had reached sexual maturity to the point where his male hormone level, or more importantly the pheromones (sex smells) that this caused him to produce, were too challenging for Blackie who then felt that this intruder should be expelled.

Not all kittens are acceptable from day one so it is important to introduce them carefully to the home and resident cat. Yet again an indoor pen is invaluable for this. The kitten can be housed inside with a warm bed, litter tray, food and water and will be protected from any initial effort from the resident cat to drive it away. The resident cat(s) and dog(s) can then grow used to the presence of the new arrival and to his or her smell within their home base. Ideally, kitten and pen are moved from room to room over the course of a week to establish a right of presence throughout the house. Damp contents from its litter tray can also be placed in the original cat's trays to help in the smell sharing process, and the cats can be fed alongside each other separated by the bars to encourage the sharing of personal space. The first free introductions should be carefully supervised with one or both restrained by hand or on harnesses. Usually the kitten, having come from an environment of sociability with its mother and littermates, will be quite unafraid of the residents' approaches and sniffs and so will not fuel any adverse interest by running away. Indeed when using this technique to introduce my own Siamese kitten, Scribble, to the resident adult male cat, Bullet, and three large dogs, it very quickly became apparent that the cage was for their protection, not hers. From day one of freedom she occupied the nine stone Bullmastiff's bed and turfs her off with a stare if ever the dog has the audacity to be in it when madam feels like a rest! The same techniques can be employed to introduce adult cats to each other, though the process may take longer. Some go as far as rubbing used litter from one cat onto the fur of the other, and it seems that this can help in some cases, if the owners can bear it. Use rubber gloves however!

Bertie's reactions to the new kitten are excessively aggressive; because of his immediate over-reaction there is little chance of him learning to interpret the kitten's arrival as non-threatening. The attacks are obviously designed to force him to leave, yet the mounting behaviour is also far from the amorous gesture we might imagine. Sexual postures are not always what they seem and in this case they are a clear attempt by Bertie to dominate the kitten in the same way as more social animals such as cows, rabbits and dogs use them to define dominance orders. Also the areas of the brain which activate during sexual arousal are close to, and in some cases the same as, those involved with certain types of arousal during aggressive behaviour. The overall reaction will clearly frighten the young kitten and as is Bertie's aim, she will be unwilling to stay around. Sex does not actually play a role in this behaviour, though it is interesting to note that sexual responses such as mounting do readily occur in dogs, and occasionally in cats when already aggressive or excited for other reasons. Mild sedatives, homoeopathic or traditional, for the established resident have been known to ease the acceptance of new arrivals where pronounced and enduring aggressive reactions are encountered, and also for new adult arrivals particularly upset and defensively aggressive towards the enquiring sniffs of the resident.

However, there will always be cats which will not accept the presence of others at all and respond aggressively at every sighting. This prospect increases with age but decreases with the number of cats owned. There appears to be a dampening effect on social aggression when the would-be aggressor is surrounded by familiar cats. To respond aggressively to any newcomer is presumably to draw attention to yourself and risk that movement- or excitement-elicited attack yourself. Hence even in overcrowded cat pens in animal shelters with apparently little room, newcomers may be accepted with no more than a sniff and a small hiss, but a big readjustment of each cat's position as they all move round to maintain as large a personal space (flight distance) around themselves as possible is likely.

With pet cats, it is essential to be reasonable in trying to introduce new cats, young or old. About three weeks of controlled introductions, with drug assistance for adults if required, is sufficient time to indicate whether the dust will settle and harmony is likely to be achieved. But after this period it is essential to face up to the fact that the original occupant will not be particularly happy with having to share his home with this new arrival, and that the new arrival may always live in fear of the original occupant and really never develop his or her own character to the full. In this case the rule of last in, first out, should apply, irrespective of how cute the new kitten looks, and a nice new home should be sought rather than persisting too long, especially with kittens, for whom being in a relaxed, friendly home environment while still young is crucial. Having said this I have several cases on file where, having tried my hardest to get cats to accept each other and reached the decision to rehome the latest arrival, the owner persisted for one more week or so and now the cats groom each other and sleep, eat, play and go out together perfectly happily, which just shows how variable cats can be and how much we have yet to learn about their social habits.

Territoriality

For all of us with outdoor cats, their territorial defence and lack of sociability with most other local cats is something we take for granted. While some do manage to make friends with next door's cat and some even establish very strong pair bonds with the moggy from the next block, the majority are intolerant of others in their patch when they want to use it. Neutering reduces much of the motivation for tom cats, especially, to be territorial and enables far higher densities of cats to co-exist relatively peacefully in our close packed suburban housing estates. Their ability to accept others once the need to maintain a territory that defines a level of food availability has been removed, is a pointer to the elasticity of the cats sociability. By neutering them we also remove much of the hormonally motivated obligation to carve out a kittening area by females or a territory to encompass those of several females by males, and much of the requirement for social contact for reproductive purposes, such as queens in season calling for mates, also disappears. The pressure to manage any territory is largely off, yet some cats remain great defenders of their patch. In highly populated areas it seems that there may not be enough of a patch for every one and the dominant or most territorial individual will find that he or she needs to be on constant patrol to repel all intruders.

Cats like Tyson (page 92) are typical examples, but most surrounding cats will gear their usage of his territory around his absence. When he is indoors and asleep, other cats will move through, aware of the reduced risk of expulsion from the staleness of Tyson's smell at his spraying posts, and from having learned that he is regularly absent at certain times of day and night. Tyson will probably be most anxious to patrol at the key times of dawn and dusk, the usual hunting times when most prey is active, and so as a dominant cat will have established unopposed right of access to his territory at these times. Then other cats may flee without putting up any opposition so as to afford him his rights and to avoid conflict, yet at other less crucial times expect to able to use the territory and walkways without hindrance. Tyson, or perhaps a slightly less assertive cat, may even be prepared to watch them walk through past his nose. The system has become one of time-budgeting over a limited desirable resource, in the same way as we rent a week per year of an attractive holiday condominium, and is yet another remarkable adaptive feature of the social life of the cat.

After many years of being ignored by biologists, the behaviour of the domestic cat, both pet and feral, has become the feature of much scientific interest in recent years. Dr David McDonald of the Department of Zoology at Oxford University and previously known for his marvellous studies of foxes, has been studying the social life of free-ranging farm and feral cats for a number of years. One of his students produced an enormous PhD thesis on the social interactive behaviour of feral cats which has done much to stimulate the imagination as to how and why cats communicate and what affects their social organization. Dr John Bradshaw of the University of Southampton, has also been looking at the social interactions of feral cats and the Universities Federation for Animal Welfare at Potters Bar, for whom I had the honour to be employed as research biologist for three years, have also sponsored much research into the behaviour and, where necessary, the control of feral cat colonies. The media too have started to become interested in the tiger in our living room and there have been several very interesting television documentaries on the domestic cat in recent years.

The best recent reference work of manageable proportions has, however, been that of biologist and broadcaster Roger Tabor, with his paperback, *The Wildlife of the Domestic Cat* . I would thoroughly recommend this to all who are interested in reading more about what their pet and his feral cousins get up to outdoors away from the fireside. We tend not to see his territorial activities as problems. What Thomas does outdoors is his affair and as we can't control or train him anyway, we shouldn't worry. Unless.......

Despotism

Dear Mr Neville,

Our two-year-old Burmese cat, Lurk, is a total hooligan, or at least has become one in recent months. For a neutered male he is now quite the most aggressive cat either of us have ever owned, or seen outside of a wildlife documentary though previously he was quite a normal cat. We have had endless complaints about him from all our neighbours, some of whom don't even own a cat! He not only chases and fights with all other cats outside in the neighbourhood, but he also seeks them in the safety of their own homes by climbing through open windows, cat flaps and open doors!

Once discovered, he attacks them severely leaving blood and fur strewn all over the house and pausing on occasion to eat the resident cat's dinner. If the owners are present and try to break up the fight, Lurk is quite likely to attack them, and he has also been known to attack young children playing in the garden. Our neighbours are insisting, one of them via his solicitor, that Lurk be destroyed and frankly we agree with their sentiments. Yet with us, Lurk is extremely affectionate so we are reluctant to have him destroyed if something could be done to build upon this home character. We have tried to keep him in but he paces and cries to go out, and is clearly very unhappy. Suggestions urgently please!

Yours sincerely

Carrie and James Stroud-Parker

Such widespread despotism is fortunately far from common in cats, though I should say that every case I have ever seen has concerned a neutered male Burmese of between one and two years of age. Perhaps this indicates some breed disposition to despotism though there may be many other despots of other breeds and lineage which have been destroyed for their Pol Pot approach to others rather than being referred to me. Such a constant expression of aggression towards all other cats with no respect for the security of their home base, redirected aggression towards any protective owner unwise enough to intervene and targeted aggression towards children would lead the majority of rational folk to believe that the state of Lurk's mind is severely disturbed, perhaps by some form of brain tumour. While this may be the case, it may be equally true that Lurk is basically a tough, dominant character who has learned the success of his own strength. 'Absolute power corrupts absolutely' as the saying goes, so having steadily achieved more territorial control over neighbourhood felines, Lurk has decided to exact the tax of total right of access to their shelter base and food bowls.

Treatment can be surprisingly successful but involves enormous amounts of co-operation with the locals. Essentially Lurk has to meet with failure and rejection for a few weeks with regards to his relations with other cats. First Lurk is taken to the vet for a course of progestins, feminizing hormones with a calming effect, or a longer lasting depot injection of essentially the same treatment in a stronger form. Then his patrol hours are rigorously controlled by his owners to coincide with times when all other local cats are safely shut in behind closed doors (and windows). At other times Lurk is placed in a pen outdoors where other cats can overcome their fear of him and approach with more confidence. Lurk's efforts to repel them or engage in warfare meet with failure, until gradually they achieve rights of occupancy in his territory and he comes to accept them more. Owners of the other cats enforce their pets' defence of home by hosing or throwing water at Lurk whenever he is seen in their garden and their cats are fed in a totally secure region of their house without any food being left outside that might encourage Lurk. Finally, any possible influence of diet on Lurk's behaviour is removed by providing a fresh chicken/fish diet for a month or so in short feeds to keep him coming home. Controlled excursions into the garden with Lurk on a harness and lead are also carried out, and any attempt on his part to be aggressive towards other cats is met with a jet of water in the face as a negative conditioning technique to build on what associative learning processes may be present. While this treatment, which is commonly employed in restructuring aggressive anti-social behaviour in dogs, may appear contradictory to the earlier comments about cats not usually associating such unpleasant consequences with their actions, in some cases it can help when accurately timed in treating aggressive Burmese, Siamese and other oriental breeds. Perhaps they are more dog-like than we realize! The above treatment clearly demands much from owners and neighbours, but has proved successful even where full co-operation has not been obtained. Such success has surprised me enormously, but perhaps we should view despots like Lurk more as school bullies than deranged cats. A good dose of embarrassing defeat seems most reformative!

Lurk is by no means unique in his despotism but imagine, if you will, the disruption such a character could cause in a group of cats which live permanently indoors.

Dear Mr Neville

I have to confess to being obsessed by cats, indeed I have twelve of them. They are all quite peaceful except for Boris, a neutered male tabby. He tolerates, without being friends with, all of the other eleven except for Gemmy, a spayed female. To say he hates her is an understatement, as he spends his entire life indoors stalking and attacking her. She is already rather a nervous cat and naturally is very upset by Boris. She slinks around and hides behind the sofa for long periods, and though we try to keep them apart, doors get left open sometimes and Boris is in and chasing after Gemmy before we realize. The ensuing fight frightens all the others who run like crazy in all directions, some through the doors, others out of the windows, others even go up and over us onto the dresser. It takes ages for life to return to normal after a successful attack. Is there anything we can do? Poor Gemmy has had so many wounds and trips to the vet, and really doesn't deserve this treatment from Boris.

Yours faithfully

Karen Black

I have to say that total peace between twelve cats is rather a lot to expect, yet a friend kept nine Burmese without any problem, other than a similar despotic relationship between Hugo, a tough male, and one of the females.

Despotism is observed in social animals although it is not usually the normal hierarchy or method of social organization. Normal canine society tends to be based on a linear hierarchy with the alpha figure (top animal), or one alpha figure from each sex being dominant, and followed in sequence by the others. None are equal and usually the order shifts only with the assimilation of adolescent pups into the pack, or the expulsion of older members. Such a pecking order system is observed occasionally in cat societies but usually only in high density groups when food is in short supply.

More usually the cats' system is based on tolerance and shifting status. Even a very tough dominant tabby in a feral colony may be unwilling to oust a weaker individual from a food item and will simply sit by and wait his turn.

Occasionally despots occur in both canine and feline groups where one war lord reigns supreme and all others have equal bottom rank beneath him (and it is usually a 'him'). Other animals, despite frequent attacks for little or no reason other than the violent enforcement of superiority, seem to allow such domination. Standing up to such characters may be inadvisable, but the underlings prefer to remain alongside these animal Hitlers, presumably for the protection they provide against unknown challenges. Perhaps feline and canine despots have the same mesmeric leadership qualities that attract and bind followers to them, as had Hitler, Genghis Khan and all other successful human despots.

There are benevolent dictators, and perhaps this system of government is attractive because it reduces the level of decision-making for the leader's acolytes. Fine, if you can find a benevolent dictator, as Gemmy's companions seem to have done with Boris. He presumably receives constant acknowledgement of his dominant position from them; they probably give right of way, access to the food bowl and can be moved from sunspots and comfortable chairs, simply with a stare from Boris. Not so fine if you are Gemmy, the scapegoat for all ills and used by Boris to exhibit the terrible consequences of challenging the despot's authority to all the others. I am reading too much into what may well be a simple case of incompatibility between two cats, but such problems can cause such enormous disruption to what otherwise could be a harmonious relationship between cats. Quite apart from all the others understandably panicking at the onset of Boris' attacks on Gemmy, the general level of tension in the house will be constantly high, and the risk of injury to innocent bystanders, feline and human, is high. Not only does someone have to separate the combatants, others may be scratched and torn by frightened cats trying to get out of the way.

While it may be kinder to rehome Gemmy for her own well-being, my experience is that such a move simply causes cats like Boris to select another victim to hound. Rehoming Boris isn't so easy either – would you want him with your cats? In any case he's probably quite a nice old tyrant when he's not attacking Gemmy. Owners usually have a liking for the Borises of this world and may have a sneaking suspicion, largely justified in many cases, that relations between all the others are only maintained by the iron fist from above, and that real infighting could break out on all fronts if he were to go.

As before, our aim in treatment is to allow controlled exposure between Gemmy and Boris, so that Gemmy's fear reactions or efforts to escape are denied and thus don't exacerbate his chasing instincts. He gets used to accepting her presence as he has done with the others and his attempts to chase or battle her away meet with failure. I commonly recommend the controlled use of sedatives for both sides in such socialization programmes, withdrawn steadily over four to six weeks as introductions develop. Progress is usually surprisingly good, though there are inevitably a few hiccups and temporary reversals, and though Gemmy and her ilk never become particularly confident cats they are at least safe from attack. In many cases they possibly wouldn't be any more confident elsewhere and may even be worse off and more vulnerable as solo cats, even if the owners would rehome them.

I have mentioned the adjunctive use of alternative medicine in the treatment of aggression quite frequently in this chapter, and never have I encountered more success with this approach than with treating despots. Four drops of the Bach's Flower Remedy, Vine, diluted with water and squirted into the cat's mouth via a plastic pipette two or three or times a day, if it won't be accepted in drinking water or milk, has produced some amazing results. Despots like Hugo under this treatment are still arrogant strutters, but instead of attacking their usual victim, they simply ignore them or, at worst, hiss and walk away. I wish someone could tell me why this treatment is so effective in these cases.

Throughout this chapter we have looked at the dangerous side of our pet cats, in relation to the cats' normal prey, other species, other cats and ourselves. With the exception of predatory aggression and territorial aggression towards other cats, thankfully all are rarely observed amongst the feline population and are only to be found in concentration in my case files. Two important points emerge. One is that aggressive cats should never be tangled with, as once excited to the point of fearful or assertive aggression, the usually discretionary cat is well armed to inflict serious injury on us. Secondly, he or she will also be sufficiently activated when already aggressive as not to recognize us as friends, nor interpret our intervention on his behalf as anything other than threatening. Where intervention is essential to interrupt aggressive behaviour, it must be via buckets of water or the use of startling devices such as rape alarms, never with hands or feet or by advancing into the fray. Such intervention may decide the battle but set back the winning of the war, which must be tackled more logically long-term looking at all the influences on the cat's behaviour. Thankfully, most problems are resolvable given sufficient time.

9
Spraying and Other Nasty Markings

The majority of the cases I see concern cats which mark their territory indoors. In fact you could say that cases of indoor spraying, furniture scratching and worst of all, middening, are my 'bread and butter'. Middening is the deliberate act of defecating in the open to emphasize occupancy of part of the cat's territory. This is the most extreme and unpleasant type of marking but is in the same context as all kinds of other marking behaviours which our cats perform, most of which are acceptable or even encouraged. When a cat rubs around our legs or offers its body to be stroked it is actually transferring some of its own personal scent onto us. The scent is produced in glands all over the cat's body and especially along the tail (caudal glands) in mature cats, from where it becomes dispersed more evenly over the fur when the cat grooms itself. The scent is then ready to be rubbed onto furniture, another cat, or us. This helps the cat develop a communal smell with us and its home so it can identify the strength of its occupancy at the core of its territory, and also recognize us as something of a protective mother figure. Scent producing glands are particularly concentrated under the chin, on each side of the forehead (temporal glands), and on the lips (perioral glands) which helps account for the enthusiastic head butting and facial rubbing that cats enjoy with us. Saliva is also produced to a greater level during such interactions to help spread the message.

The presentation of the underside of the chin by the excited purring cat is far more than just a display of affection between friends. It is also a way of cementing that relationship by the cat anointing us with his own scent, and it's the way friendly cats greet each other. By presenting their faces and rubbing each other they each acquire a little of each other's scent to cement their bonds and ensure that they continue to recognize each other after any short period of separation. The activity of these glands is reduced when male cats are castrated and this accounts for much of the loss of that tom-cat smell. Perhaps a cat's personal scent is also altered when it is traumatized for it could be that the recognition system suddenly breaks down so that an accepted friend is no longer identifiable, and is repulsed in the way described in the chapter 7 (pages 61 -65). We don't know, nor do we understand very much at all about the importance and method of scent communication in cats, or any other other social animals for that matter. It simply isn't our medium and though we have the sense, we don't have the senses to detect what is going on.

We can, however, see where the cat combines scent marking with a visual signal at scratch posts. Most of these posts are outdoors in the competitive zones of gardens and countryside, but occasionally cats will need to scratch indoors too. More than just keeping claws ready for action by stropping on chairs, curtains and, favourite of all, Hessian wallpaper, the cat is also leaving a definite scent mark derived from glands situated on the sides of the pads. These glands also help the cat mark out walkways indoors and out, both to identify the cat's usage of the path to other cats and to help the cat navigate its way home or to favourite activity areas or sleeping and feeding lairs.

Scratching

Dear Mr Neville,

Sidney is a lovely cat but he will insist on scratching my best furniture at the slightest opportunity. My sofa is shredded and many of my chairs are in rags. I've tried telling him off but it only discourages at the time. He simply returns later to have another go. Is there anything I can do?

Yours sincerely

Samantha Swift

The first thing we have to do with this type of problem is decide just how much of the scratching is likely to be of a scent marking function. Usually the cats which scratch a great number of areas in the house, especially around the access points near doors, are marking their territory with scent and the visual sign of the scratched surface. These must be treated in the same way as indoor sprayers and the most important feature of treatment is finding out why the cat feels the need to mark what should already be secure. Cleaning up the marking posts and controlling the cat's access to it for a while, as well improving his perception of home as a safe base are also important and we will take a closer look at this later in the chapter.

When the problem is centred on one or two scratching posts only, the emphasis is more on keeping the claws in trim. Scratching doesn't actually sharpen the points on the claws; in fact, it causes the whole outer layer of dead claw to slough off and exposes a new, pointed claw-tip underneath. It is often easy to find old claw outer layers discarded on scratching posts in this way. Cats which scratch indoors may be doing so to avoid drawing the attention of local rivals outdoors, but usually it arises through a lack of transfer, as a young cat, from using available surfaces at home to the more acceptable outdoor surfaces of trees and fenceposts. Treatment is therefore aimed at trying to redirect the cat's claw sharpening activities rather than punishing him or her for doing what is needed naturally to keep the weapons fit for hunting. Placing a suitable alternative of a type of material that is not found elsewhere in the house, on or directly in front of scratched areas usually does the trick. A large piece of bark or one of the commercially available sisal or string-wrapped posts or cards can easily be positioned at the scratching areas and once the cat has begun using it, can be moved a few inches at a time to a more convenient area nearby in the room. Using a bark alternative may even help the cat transfer all scratching to the outdoors.

The need to scratch is obviously inherent in all cats as part of keeping in shape and so a suitable scratch post should always be made available to young kittens which are to be kept indoors permanently. This is far more sensible than the outrageous American attitude to solving the problem by surgically removing the cat's claws. Thankfully, few veterinary surgeons in the United Kingdom are prepared to carry out this unnecessary mutilation except on medical grounds and especially not if the cat is ever to be allowed outside where it would need its claws for climbing, hunting and self defence. While the adult declawed cats I have seen seem to bear no ill-effects long after the surgery, I must say that the most miserable, healthy cat I ever saw was one in an American Animal Welfare Hospital which had been declawed the day before. I only wish veterinary surgeons worldwide would take a strong stand against all cosmetic mutilation of pets, including the docking of dog's tails.

Spraying

Most people do resolve problems of scratching indoors by providing posts or by the cat learning that only one end of one old chair is acceptable. The majority of the cases of unwanted indoor marking that are referred to me concern the spraying cat. Spraying is a very deliberate act of marking by males and females and not to be confused with urinating, which is more usually performed from a squatting posture into a prepared hole in a movable substance such as soil or cat litter. Spraying is performed from a standing position, with the cat facing away from the feature to be marked. Typically the cat stands a few centimetres away from a vertical object such as a fence post or shrub and with tail held high and quivering at the tip, directs up to 2 millilitres of urine backwards in short spurts, designed to spread the spray over the target area. The process looks quite difficult to perform and perhaps the necessary muscular spasms around the penis in males account for the stepping action of the hind legs, the slight arching of the back and the heavily concentrated look on the cat's face.

Scent marking is performed by many animals and often readily detectable by even our poor sense of smell. Polecats, mink and ferrets, dogs and mice all deposit very distinctive odours in and around their territory, but there is little to compete with the unforgettable smell of tom cat spray. Such smells are known as pheromones which are secreted in a fatty viscous form from glands placed all over the body. Pheromones can simply be released into the air, particularly by social animals, but are specifically directed to concentrate scent on marking posts by many mammals. Glands producing pheromones are found on the abdomen, feet and face of of mice and many species of deer mark grasses or shrubs with secretions from glands below the eye. Many mammals have their most active scent glands located in the genital region. Urine and faeces are good vectors for distributing pheromones and so are anointed with smelly secretions from the accessory glands either during normal elimination (scientific hype for toileting) or by specifically targeted application as in spraying cats or leg-cocking dogs.

While insects such as bees and ants also communicate by pheremones, birds do not, and though many smell, they do not seem to use this as a method of influencing each other's behaviour. With cats, scent posts are used to allow a longer lasting signal to be given rather than simply releasing smell into the air. The signal may last hours or even weeks after its deposit and be detectable, given the right atmospheric conditions, up to fifty feet away. The importance of smell communication is easily overlooked, as the cat outwardly appears to be a creature evolved to rely on senses of sight and hearing; certainly these senses are employed most in a cat's relationship with us.

Males of scent-marking species usually possess more scent glands than females and mark more frequently. Cats are no exception and un-neutered males spray more often, and certainly more pungently than other cats. Few people can live with an entire tom (a lovely expression for the uncastrated male) who sprays indoors, which is the reason why males are usually castrated before the behaviour develops at adolescence unless the lucky chap is to be a stud. In this case he will probably have to accept life in his own personal enclosure at the bottom of the garden, with the compensation of occasional trips to the cat show in order to be admired, win prizes and be signed up to perform with a series of attractive young female cats. Even if you smell, life can have its compensations when you're a cat.

Entire toms begin to spray at adolescence with the onset of male hormone production, as the result of the need, it is thought, to become socially active in procuring a territory and attracting females. Spraying frequency, together with the rise in testosterone levels, increases with the presence of females in season and during the feral or stud feline breeding season. Consequently, the increased movement of toms means there is a growth in the number of competitive encounters with other males. Interestingly, castration after the cat has started to spray causes an immediate cessation of the behaviour in nearly eighty per cent of males of all ages, and slower diminution of the behaviour over a few months in another ten per cent. Roaming and inter-male fighting also decrease in the majority of cases. A few males continue to spray despite being emasculated, and a few which are castrated before puberty also start to spray some time later. This indicates that while spraying is largely a secondary sexual behaviour related to the presence of androgens, specifically testosterone, in the mature tom, other factors can also cause them to spray.

The smell and positioning of the spray at nose-height to other cats helps the sprayer announce his presence to existing users of the garden or other territory. The spray is believed to act as a personal calling card, giving the finder all sorts of details about the sprayer. It probably indicates sex, rank and sexual status, state of health, age, and the freshness of the mark will indicate how long ago the sprayer was at that point. A fresh mark may mean 'beware' while a stale one may mean 'relax' and proceed in safety. Yet many cats simply investigate a sprayed area and walk away apparently unconcerned, without any modification of their behaviour. They may over-spray the mark, or they may not, and there seems to be little pattern in their response to the spray of an observed highly territorial individual, from one day to the next. Perhaps spraying is like a present, more important in the giving than the receiving. Tigers, especially, spray extremely frequently if placed in an unfamiliar area perhaps in an effort to surround themselves with their own comforting familiar smell, rather than it being an announcement to others of their arrival. Feral toms too may enter a strange area where food is provided regularly and when the area is quiet and no actual food is available, spray close by to establish their smell as an envoy to resident cats. This is believed to desensitize the residents a little to his actual physical arrival some time later when the food appears. His arrival is thus staggered and as his physical entry is less of a sudden alarming prospect to the resident cats, he is perhaps less likely to be ejected *sine die*.

However, it is hard to imagine that spraying is not intended to have some deterrent properties for rivals trying to compete for food, shelter space and especially females. Hyenas have been observed to chase prey (they do hunt as well as eat carrion) as far as the boundary to their territory. Once the quarry crosses into that of the neighbouring pack, the hyenas stop abruptly rather than violate the border agreements. Instead they mark the border with urine, faeces and its associated pheremones, and withdraw into their own area again. Frequent inspection of the border marks would reveal whether the rival pack had marked recently and was still in residence or had moved on, in which case they may feel safe to move in if desired. Generally mammals do not just mark the borders to their territory, but it appears that way with cats because fences are places where progress is slowed and encounters with other cats are more likely. Instead, cats, and hyenas, mark throughout their territory, perhaps endorsing the idea of the self-familiarization function of spraying observed in tigers. The more marks, especially fresh ones, that are encountered in a given area, then the more the sprayer feels confident within it, but others moving through need not necessarily feel ill at ease. With the high density of pet cats in suburban and urban areas spraying posts are constantly marked and over-marked by a sequence of cats, depending on who is moving through the area and at what time. Doubtless the sprayed posts help define the time-budgeting system of territory occupation described in the previous chapter.

For any form of marking to be effective, the message has to be understood by the receiver . Observing cats' responses to their own marks and especially those of other cats is enormously interesting because it involves a sense which we do not possess. The flehmen grimace of the cat is also observed in other animals such as horses and cattle which possess the vomero-nasal, or Jacobson's organ. We, like all other primates, do not possess this organ and so lack this sense. When sniffing spray posts or urine the cat may seem to be in something of a trance and with ears slightly flattened, it wrinkles its nose and lifts its lips while drawing in the scent. The cat's tongue may flutter a little and breathing may stop for a short while as the scent is drawn in through the gaping mouth and concentrated in the vomero-nasal organ, which is situated at the base of the nasal cavity and open to the mouth via two ducts in the hard upper palate behind the front teeth. The sense is therefore best described as mid-way between taste and smell and enables the cat to 'taste' the full message of the smell under investigation. Sex smells elicit the flehmen response best, especially in entire toms, as does the smell of cat nip with those cats which are sensitive or responsive to it.

So cats smell and 'smell-taste' the messages left by others, and to check on the freshness and content of their own marks. Just how much information they can gather this way is a mystery to us, but clearly of vital importance to the social life of the cat. Some American behaviourists have even suggested that destroying the cat's sense of smell surgically by means of an 'olfactory tractotomy' or less drastically, by blocking the ducts leading to the vomero-nasal organ, will decrease the frequency of unwanted spraying in cats. Unable to detect the decay of its own smell and with smell playing a far less important role in the indoor sprayer's overall sensory input this may be a logical, if totally unacceptable approach to tackling such problems. I would never recommend such surgery and would always rely on identifying the cause to the problem and/or managing the cat's access to that influence so that it learns to cope without spraying rather than interfering with a major natural sense. Sometimes I wonder what right we have to go chopping animals around in an effort to modify behaviour problems. Castration, yes, both for our benefit and the cat's in most cases, but removing the sense of smell? Surely not.

Contrary to popular belief, it is not only entire males which spray. Entire females may spray when coming into season to announce their receptiveness to nearby males. They appear less likely to spray at other times and one would expect them not to spray close to the nest when nursing kittens to avoid attracting attention. But neutered cats, male and female, also spray their outdoor territories as a normal pattern of behaviour. Many owners seem surprised at this and feel that perhaps their female cat is metamorphosing into a male, or that their cat, male or female, is having difficulties in urinating and is being forced to adopt a standing posture for the purpose. The vet usually assures them that all is well, though if a cat looks in pain or discomfort when spraying outdoors, or starts to spray indoors, it is always wise to ask the vet to examine the cat physically and carry out a urine analysis. Cats of both sexes can adopt a standing posture to urinate when suffering from cystitis, an inflammation of the bladder caused by injury or infection, or if suffering from a painful condition known as feline urological syndrome, or FUS for short. In this syndrome, uroliths or 'stones' may become deposited in the kidneys or ureters causing partial or even total blockage of the passage of urine. The syndrome was a frequently encountered problem when dry feline diets were originally introduced some years ago. The diets contained too high a level of magnesium as well as causing an alkaline urine , creating the ideal environment for precipitation of mineral crystals in the urinary system. Thankfully the formula of the dry diets was quickly altered and one or two companies investigated the problem extensively before launching safer diets. However, as such problems may not just arise as a result of unbalanced diets, any sudden onset of spraying behaviour in a previously non-spraying cat or difficulties at the litter tray should prompt an immediate visit to the vet. Caught early, FUS can usually be treated and involves close dietary

114

management as much as anything else to prevent recurrence.

Virtually all cats, including neutered males and females, spray outdoors as a normal part of their behavioural repertoire. The higher the density of cats, the more they spray, and this is especially true of spayed females outdoors. As well as fighting more, assertive un-neutered toms also spray more than other cats because territorial occupancy is most important to them, and so the need to impress a chemical presence on the opposition is perhaps linked to aggression. But neutered cats also spray more when crowded, without necessarily becoming more aggressive. Therefore, instead of viewing spraying as a dominant supportive gesture in maintaining rights to certain resources, we should perhaps view it as an emotional response to certain conditions or changes which includes, at the extreme end, arousal when aggressive or territorial.

Cats of both sexes vary enormously in their propensity for spraying, and the breeds which we recognize as more sensitive or intelligent, such as many of the orientals, seem especially practised at the art. Cats may spray more when they are anxious or when vulnerable, perhaps because of a decreased level of attention from their owners. They may spray when their routines with us are disrupted, or if they are punished, or forced to share our affections with a new cat. For others spraying may be triggered by a combination of influences raising the cat's emotional state beyond a key threshold which is relieved partially by the act of spraying to surround the cat with a familiar security veil of smell. This includes those times when a cat may be aroused through the anxiety of defeat or else being victoriously assertive following an encounter with a rival. The spraying reaction threshold is apparently higher in neutered cats than un-neutered, and highest of all in spayed females. In short, the non-reproducing female is perhaps the most competent, least emotionally reactive cat compared with macho toms, sensitive neutered males or unspayed females whose behaviour is subject to a varying hormonal status. Certainly most of the cases of unwanted spraying behaviour referred to me concern neutered males. It is also true, from my experiences and those of behaviourists in the USA, that while spayed females are less likely to be presented for treatment of a spraying problem, when they are, they are far more difficult to reform than males, neutered or un-neutered.

Thus spraying is a natural behaviour in all feline species and, for whatever reason or combination of reasons, has persisted because it bestows some evolutionary advantage on sprayers compared with non-sprayers. Or is it that more reactive, emotional cats, which are more likely to spray, are also more likely to sense and avoid danger and so survive better? Whichever the reason, spraying is only perceived as a behaviour problem in pet cats if it occurs in our homes, disregarding as largely untreatable the anointing of our garden shrubs or back doorways by free-ranging feral cats or unidentified local pets. Fortunately the majority of our pet cats do not spray indoors, because the core to their territory is secure from challenge by other cats and our pets are most likely to be emotionally calm there. We are there to protect them and behave like their mother, and they can rest without fear of disturbance. In the territorial sense it would also be a waste of time and energy to mark an area already defined as occupied and safe. In cases where many cats share a household, even apparently reactive or victimized individuals may not spray to avoid drawing attention to themselves, and so enable them to remain in the home without being perceived as unnecessarily weak or challenging by other more dominant house-mates.

So when a cat does start spraying indoors, it is usually an indication that the cat has been upset or challenged by some disturbance or change in our home which has made it feel less secure. Identify the challenge and remove it, and the cat's emotional state may wind down below the spraying threshold. The difficulty is usually the identification part, as we may be dealing with a cumulative effect from several influences, each one of which may seem relatively minor. The other difficulty is that the emotional threshold of every individual cat is different and while we can identify certain gross triggers which would set any cat off on a spraying career, others seem enormously sensitive and start spraying in response to the most minor of changes. These are the most difficult to treat, as aspects of treatment itself may be sufficiently worrying to cause an initial upsurge in the frequency of spraying. Then there is the problem of not being able to catch the cat in the act of spraying. When we are with the cat we act as a protecting influence, an emotional prop that allows the cat to settle and feel confident to the point of not needing to spray. So it may be difficult to identify when and under what circumstances the cat will spray his marking posts or to organize treatment or manage access to those points on anything other than a total exclusion basis.

If there are lots of cats in the house it may also be impossible to identify the culprit or culprits if they are never caught in the act. In which case we may need to inject each cat subcutaneously in turn with a small amount of a liquid called fluorescin. This causes the cat's urine to fluoresce under the ultra-violet light of a Wood's lamp, which is more often used to detect ringworm. Fluorescence of fresh spray will indicate that the cat injected on that day is the target for treatment and all others *may* be innocent until tested. And if that isn't enough for a poor old cat shrink, there are also those cats who cope without spraying after enormous disruptions such as being beaten to pieces by the local feral un-neutered tom on one day, and then anoint every piece of furniture after being dive-bombed by a bird the next. Delicate souls, cats! Treat each one with individual care and attention!

Dear Mr Neville,

We have enjoyed six trouble-free years with Bandit, our neutered male cat. He has followed us through three moves of house and always established himself well with local cats. He is not a great fighter but usually stands up for himself okay. Three weeks ago we came downstairs one morning to find that overnight our kitchen had been a feline battlefield. Another cat must have come in through the cat flap to engage Bandit. We must have slept right through an enormous dust-up as there was fur everywhere. Most of it was Bandit coloured so we presume that he got the worst of it. He bore a few scars but seemed unaffected in his general behaviour, except for the fact that he now sprays all over the house. Not a piece of furniture has been left unsprayed, and he regularly makes the rounds and tops up his marks. This only happens when we are away so he obviously knows that it is wrong. Should we punish him? Our house is starting to smell most unpleasant despite hours of cleaning up after him. Please help!

Yours sincerely

Sue and Barry Scott

If there is one single most popular invention for cat owners, then it must be the cat flap. Believed to have been developed by the great physicist Isaac Newton, who, between having apples drop on his head, designed a small cat-size hole in his door so that his cat could come and go as it pleased without interrupting him. Had he but paused a little he would have realized that holes and, later, simple cat flaps with hinged doors, don't just allow the resident cat to go in and out. All his friends and enemies can come in and out too! Some pets are surprisingly tolerant of other cats coming into the very core of their territory, stealing their food and resting in their favourite chair. Others never suffer such invasion, perhaps because local rivals have never experienced a cat flap and have no idea what joys lie beyond them. Then there are those like Bandit, for whom home is always a castle, wherever it is. Here he is protected by the family, loved and cared for and always has those safe four walls to retire behind away from holding his own against the local rivals. However, the introduction of the simple flap has meant that the drawbridge is always down and any other cat can also come in. If it happens to be the local un-neutered tough feral tom, or the most hated of rivals from two houses away, we have unwittingly created the most inappropriate battle-zone. Food, shelter and affection, even from sleeping and non-protecting owners are resources worthy of spirited defence.

Although ultimately Bandit managed to repel the intruder, this incursion into the safe core of his territory has meant that Bandit now perceives the home as being under the same challenge as the outdoor section of his patch. As other cats can now enter at will, Bandit must try to endorse his presence and occupancy chemically, by spraying all areas likely to be encountered by any other cat. Spraying will also surround Bandit with his own familiar smell, which is perhaps now more of a comfort than the communal smell of home, owners and perhaps the other pets that he shares with. Since the invasion this shared smell is no longer sufficient to make him feel at home, and he must now rely on his own efforts. Emotionally, this is about the worst thing that can happen to a normal cat. Small wonder that he sprays everything, and tries to ensure that all marks are kept fresh. At first, as the owners clean up, the cat goes round re-anointing the marking areas. Long after the original fight, and even if it only occurred once, a cat can be sufficiently disturbed as to keep on spraying indoors for months afterwards.

If this sounds a logical reason for a cat to start spraying, there are plenty of examples in my files of cats which have begun spraying indoors simply after the installation of a cat flap. The owners felt that they were doing the cat and themselves a favour, but all they did was to destroy the important barrier between the outdoor competitive jungle and the safe indoor reserved zone. In some cases the cat immediately recognizes that the whole of his territory has become continuous and that there is no safe castle anymore. So if spraying is practised and helps outdoors, then it ought to help indoors too. The cat is of course unable to distinguish between apple trees in the garden and antique Chippendale chairs in the lounge. After all they are both vertical and will hold the smell of his spray at exactly the right height.

It doesn't take a huge amount of intelligence to deduce that cat flaps are bad news for cats like Bandit, and that if the indoor spraying is to cease, the flap will have to be boarded up again so that security of the inner sanctum may be re-established. Even so, the cat may continue to spray for a while and will need emotional mollification through extra love and affection from the family, and perhaps a little short-term sedative or progestin treatment from the vet, as suggested for treating some cases of feline nervousness.

Cleaning

It is just as important to clean up the indoor marking posts effectively so that Bandit and his ilk do not go back to them to top up afresh over the decaying or stale smell of previous efforts. This alone can be a sufficient stimulus to prompt spraying, even though the original cause of the onset of the problem may be long gone. Unfortunately cats and people have different ideas as to what smells clean. To us the sterile wafts of chlorine or light ammonia wafts denote that the area is as clean as it can be, and we buy our household bleaches and many detergents on this premise. Sadly, ammonia and chlorine compounds are constituents of cat urine. To us the freshly scrubbed spraying post of the table leg smells as sterile as a Swedish kitchen. To the cat, it probably smells as if another cat has overmarked his mark. In the face of such a challenge, the cat will simply over-spray the over mark and so on ad infinitum!

Therefore, when cleaning sprayed areas it is vital to consider the cat's nose if we are to avoid further problems. Ideally the smell of the spray must be removed entirely without being substituted with a compound that will itself leave a residue containing ammonia or chlorine, however clean it appears to our clumsy olfactory senses.

Ideally too, the smell of the cat's spray should not be replaced with the artificial smell of lemons, pine, roses or any other of the standard scents added to cleaners and air fresheners. These scents will attract the cat back to his spraying post and inform him that something has been and found his mark sufficiently interesting to warrant leaving its own calling card. It may not be another cat, but it is something unencountered previously and clearly is presenting a challenge. Time to overspray again! Cleaning sprayed areas effectively to the cat's nose is not an easy task but seems to be best achieved by using a warm ten per cent solution of an enzymatic or biological washing liquid (or powder) to wash the area initially. This removes proteinaceous compounds in the spray. Then the area should be rinsed in cold water and wiped dry, and then sprayed with a low grade alcohol such as surgical spirit via a plant mister. The spirit should be lightly agitated into fabric or wooden areas with a nail-brush so that it penetrates the fibres or minute crevices where the spray may have reached. The spirit will dislodge any fatty residues that the washing liquid couldn't dissolve. The smell of the decay of these fatty residues in particular may prompt the cat to top up his own marks, so it is essential to try and remove them. A word of warning, however. This cleaning system may also remove dye from carpets, curtains and other fabrics, and polish or varnish from furniture, so check before wading in by testing on an unimportant or hidden corner of the surface first. After cleaning, the area must be allowed to dry thoroughly, if necessary using a hair drier on very wet areas, before the cat is allowed anywhere near his old spray post.

This cleaning system is usually effective against even the particularly pungent fatty spray of entire toms, depending on the accessibility of the cleaners to the sprayed areas. Alas, some spray will often run down the vertical area it was intended for, or seep deep into soft chairs etc. and there is no other option for removing the smell but to rip up carpets or throw away chairs. My own clients' experiences indicate even the fruitlessness of putting chairs sprayed by entire toms into storage in the garage for the smell to wear off. It doesn't. But spray from neutered males and females is usually removable using the above method. It is a little long and tortuous and so I have been working on a single bottle approach with one of the leading veterinary hygiene companies, Vet Health. 'Katastrophe' should have equally effective properties in dealing with cat spray and leave no residual smell, as well as killing any harmful infectious agents in the spray. This is an area where Vet Health have unrivalled expertise. Katastrophe should be available to suffering owners of sprayed houses in early 1990.

Deterrents

Removing the smell of spray indoors will at least help prevent 'topping-up' spraying, but if the cat is still distressed or the cause of the distress has not altered, then he will still be motivated to spray his marking posts. The same spray posts will continue to attract him or her because of their important geographical position in the house. Access points such as doorways, and boundaries to the house such as windows are recognized as weak spots in the castle and deserving of extra identification. Any rival which gains access is most likely to encounter the resident's smell at these points, and also on table legs and furniture which are close to the walkways a new cat might follow in the house.

So the aim is to try to deter the cat from spraying those frontier posts of curtains, window frames, doorways and crucially placed table legs. There are countless products on the market said to deter cats from soiling indoors or in the garden. Like the more traditional use of pepper dust, chilli powder and mustard, they are usually unsuccessful. We are, after all, dealing with an emotionally disturbed cat and to receive a nose full of pepper, chilli or evil smelling chemical preparation at one's spraying post's is probably only going to upset the cat more, and cause an increase in spraying. While some claim success tying aluminium foil strips to spraying posts on the premise that the cat is startled by the noise of the spray spattering on the foil, I have never found this to be effective. I have had some success by placing unpleasant walking surfaces, such as trays of marbles or pine cones, at the base of spraying posts as while standing on a movable surface the cat cannot adopt the spraying posture. However, in most cases he simply stands a bit further away and sprays harder, or moves along a little to another point. In other words, such tactics do not block the motivation to spray, they simply re-direct it. Others in America have tried using remote punishment techniques of finely set mousetraps which go off as the cat goes towards the area and vibrates the trap with his footsteps; while some have used an electrified foil wrapping around the spray post which when sprayed upon causes a small, startling current to run back up the spray and connect on the cat's rear end! Not nice, and not even effective; as with pepper dust and all those smell or taste deterrents, the most that can be achieved is to transfer the spraying behaviour elsewhere as they do nothing to alleviate or recondition the actual motivation to spray. As with aggression problems, punishment has precious little effect on feline behaviour other than to cause flight.

Food on the other hand is usually an excellent deterrent to spraying, though it should be said that some individuals do simply move on to non-baited areas. I have yet to encounter the cat which will spray its own food presumably because it would make it distasteful to consume later. I say 'presumably' because tigers are known to spray uneaten prey having eaten their fill at a first sitting. They scratch vegetation and soil over the carcass, and then spray it, perhaps to deter others from showing interest. The behaviour has probably evolved in tigers because although solo predators they are capable of bringing down prey larger than themselves, and certainly more than can be consumed at a single sitting. There may be enough left on the antelope joint to provide a cold meal the next day, so the covering and identification is perhaps an act of lardering that is unnecessary with smaller felids which have evolved to catch only single meal-size prey.

This is just as well, because food placed on a saucer or in a little plastic container and placed at the spraying post will usually deter the indoor spraying cat. Dry food is recommended for the purpose as it lasts longer, though it may need to be stuck down on the saucer to prevent the cat from eating it. Some cats I have known have emptied the saucer and then sprayed the post! Also the family dog, for whom stealing the cat's food is one of life's main purposes, may render the method quickly inoperative unless the food is stuck to a heavy bowl which is in turn stuck to the floor. Dry cat food seems to function as a deterrent to spraying even if the cat usually receives a canned diet and rejects dry food as part of his normal ration. The reasoning behind this is obviously not that the cat finds food unpleasant, as he would a nose full of pepper dust or an electric discharge to his nether regions, but that food itself is a reassuring influence in the core of his territory. The core corresponds to his feeding lair, and perhaps wider dispersal of food in the area for a short period helps re-establish that core as being a secure feeding home-base. Similarly, flooding the sprayed rooms with objects familiar to the cat, such as toys, litter trays, beds etc. helps to reassure and negate the need to spray to achieve the same result.

Dear Mr Neville,

After eight years of sitting in the same chairs we had when we were married, my husband and I decided to splash out on a new suite. We also redecorated our living room to match the new suite, but ever since we sat back to enjoy the new ambience of No. 28 Redford Road, the cat has sprayed our new suite daily. We have been forced to rescue the plastic covers from the dustbin to cover the suite and protect it from Mick, who has also registered his disapproval of our choice of wallpaper by spraying that too in a couple of places by the door. We may have to disagree with Mick on taste and colour but how do we stop him spraying his opinions?

Yours sincerely

Moira and Arthur Alexander

This type of spraying problem can hit hard at any family. To them they have improved life for everyone, including the cat. He now has a bright airy room and new chairs to spread himself over if he's lucky. To the cat, the very heart of his territory has been totally altered. Furniture has moved and all his walkways and escape routes are blocked, or simply not there any more, if a new carpet has been laid as well. At cat's eye-level the whole scene is unfamiliar, and it smells strange too. Somehow the territory has been altered completely and filled with novel objects, so to reassure himself that this is indeed home and to ensure that no other rival moves in to try and occupy this virgin resource, many cats will spray new furniture and wallpaper for a couple of weeks. Some are so sensitive to change within their lair that they will be disturbed simply by moving existing furniture around or adding a single new fixture. Indeed, many will routinely spray any new item brought into the home, from black plastic rubbish bags (always a favourite target) to shopping baskets, kettles and the childrens' new toys. Some spray each new thing once only, but others persist for months in their daily routine of identifying all recently acquired chattels. It can take ages to make the rounds, not to mention an enormous volume of spray. Other favourite targets are those where the cat's spray becomes altered after deposition, perhaps by warming. The casual squirt on the washing machine may even go un-noticed to us in a kitchen smelling anyway of food, wet socks and babies. Un-noticed that is, until the washing machine is turned on, warms up and pervades the house with the unmistakable aroma of cat spray. The cat will notice even quicker, and as the chemistry of the smell is probably altered a little by being heated, he may think that that invisible rival has been in again and over-marked his mark.

Clearly most cats are not so easily disturbed by the arrival of new items in our homes, nor by the process of redecorating. They are often used to a constant barrage of novelty within our homes and after a simple sniff and exploration of the items or a quick tour of the newly arranged room, are sufficiently re-assured that all is as it was and home is still home. A few are more easily upset, particularly those which live permanently indoors and are unused to a frequently changing lifestyle full of daily challenges and decision-making. For them, new furniture or objects, or even the arrival of visitors can be a major event, and perhaps sufficiently disrupting or threatening to send them over that emotional threshold of insecurity.

For such sensitive characters, reducing the size of the territory to one, unchanging and secure room or pen for a while, may help the cat feel less emotionally upset. From there it can slowly be introduced to other rooms in the house with the owners acting as security bridges at initial exposure. Rooms which have been altered or which are subject to continuing change as, for example, when we build extensions to the kitchen or living room, should only be available to the cat when they are as complete and unchanging as possible. Such areas should be explored last of all following an outbreak of spraying. It should be said that living with such cats can mean that while life is calm there is no spraying, but any minor change to the home or routine could cause the problem to start again. So if you know change is coming, it may be wiser to board the cat until it's all over, or keep a supply of sedatives or other suitable treatments in the cupboard so that treatment can be administered to the cat and his reactions dulled a little before the onset of the changes.

Dear Mr Neville,

I have owned three neutered cats, two girls and a boy, for over three years with never a problem. Two weeks ago a friend asked me to take on a homeless young kitten for a few days while she found it a home. Needless to say, I can't bring myself to part with the kitten now, even though he appears to have caused something of a disruption for the others. George, the quietest character of the original three, has begun to spray all over the house since Carla arrived. I would have thought he would have accepted a female kitten without any problems, especially being used to sharing his patch with two others already. What can I do?

Yours sincerely

Janice Wooley

If the single invasion of a rival through a cat flap, or even the possibility of it, is sufficient to set some cats spraying, it is not surprising that the arrival of a new cat permanently in the house will set others off. While early experiences of social living will usually improve any cats ability to tolerate others sharing the home later, there does seem to be an individual limit for each cat. Perhaps it is more due to loss of personal space than competition for resources, but while some cats like George start spraying as soon as the least threatening form of feline challenge is presented, others live quite happily in groups of fifty or more with not a sprayer to be found. Indeed a wonderful character of my acquaintance named Bernadette keeps over one hundred and forty cats in a three-bedroom semi and an outside run in Harrow and without a sprayer among them. More than this, Bernadette is constantly introducing new cats and temporary lodgers as part of her excellent cat rescue activities. The one hundred and forty two residents (correct at time of going to press) are all neutered, exceptionally well-cared for and very harmonious in their relations with each other. It is quite a remarkable sight and perhaps leads one to believe that there is a threshold density of cats, beyond which none will spray to try and maintain an impossible hope of personal space. This does not necessarily imply that they are suffering in any way as all Bernadette's cats eat well, play and show no behavioural vices. Yet beneath this threshold, it seems that the more cats one acquires, the more likely it is that at least one will spray. I can offer no guideline to owners of ten cats with a sprayer or two in their number as to how many more to take in until the compression threshold is likely to be reached!

With regards to sensitive cats like George, who are just below the threshold of spraying while living with two others, it is essential that any new cat is introduced very carefully indeed. A new kitten or cat should be housed in a pen for a while in the same manner as described in the chapter 5, for in this type of case we are considering a cat which cannot cope. Outwardly it may show no other signs of fear or aggression, but the spraying is a strong indication that the cat cannot maintain its personal space to the point of feeling confident. Slow introductions, with the new arrival confined to the pen and only allowed controlled access to the rest of the house one room at a time, will perhaps help George habituate to the new cat's presence, but more often than not he will need a little help from the vet for a few weeks. A low and decreasing dose of progestins help cats like George to acclimatize slowly to the presence of a new cat and to develop a lower expectation of space and self-determination in the home. Treatment should be coupled with cleaning up and the use of food deterrents as outlined earlier.

Interestingly, neutered males seem most likely to be upset and spray at the arrival of a new cat in the house, especially if they already share with a female cat rather than other males. Perhaps the females are regarded as a resource and even the neutered male is simply defending or protecting his harem in such cases. Neutered males do however respond well to progestin treatment in most cases. Unneutered females may be most intolerant, perhaps because of the need for a larger personal area to allow for growth and development of kittens. Spayed females are more tolerant and seem to be more accepting than neutered males. But if they do react and start to spray, it is my experience and that of American behaviourists, that progestin treatment is not always effective at stopping the cat spraying after the period of prescription. In short, we can calm the cat's emotions so that it tolerates a new arrival and does not spray while the drugs are effective, but there is no learning or habituation process established to prevent the cat spraying after the course of drugs is complete. In such cases it may be kinder and cleaner to rehome the new arrival as soon as possible.

Dear Mr Neville,

My Siamese cat Mao is a very loud attention-demanding

126

dog! When he is active he expects the world to amuse him and for me and my family to act as his slaves. If he yells for food we feed him. If he wants to go out, we get up and open the door and if he wants affection, we drop whatever we're doing to cuddle him. If we don't, he sprays us, so rather than be wet all day, we pander to his every whim. Is there a way out of this without upsetting Mao? He's even started to spray non-cooperative visitors as well!

Your obedient servant (of course)

Margaret Thornley

This is a typical example of the learned sprayer. Mao, being a brighter-than-average member of a brighter-than-average breed has simply learned that he can obtain his owners' attention or anything else by spraying. If rubbing round the legs and purring didn't produce the desired results one day, his frustration or emotional upset at being ignored may have caused him to lift the tail. The owners, expecting the worst, immediately diverted all attention on to Mao, and the first stage of the learning process was established. It sounds as if he has even learnt to allow a short delay between yelling or posing for the owners to notice and respond appropriately, and withhold the spraying if the favourable response is received. The tail is doubtless held up straight, and quivering slightly at the tip, to indicate the real consequence of failure to respond. Mao and his type will wait a short time, and if the desired results are not achieved, the spraying ensures that at least some attention will head his way. It may arrive in the form of a shout and a size twelve boot, but it's better than being ignored and is still a form of success to the demand. The owners, of course, progressively train themselves to respond ever faster to the cat's desires, and the cat becomes a focus of constant attention so that his wishes are not inadvertently ignored.

Treatment inevitably involves a little role reversal, with the owners re-establishing themselves as the instigators of relations with Mao, and not the responders. This may mean a period of heightened spraying for a while, but it is essential that Mao's demands cease to be rewarded at any time. He has to fed, fussed, let out and totally managed at the owner's whim, not his own, and none of his threats or actual spraying should ever be responded to at the time. Cleaning up takes place when he is elsewhere. Mao has to be ignored, though if he hasn't become less of a reactionary after a few days, use of water pistols or loud distracting noises as negative reactions to his threats can help recondition him. Such tactics have to be seen as remote punishments however, and not as punishments from the owners. Otherwise this too may be seen by Mao as success in his effort to procure attention. Such conditioning can only work with the more intelligent cats which have already learnt such sequences of behaviour. In essence, cats like Mao are being treated much more like dogs, and the techniques of restructuring learned or uncontrolled behaviour will be much the same in both species. Usually only a short period of restructuring is necessary, and is simply a case of making one's your mind to do it and living through the consequences of owning a confused cat for a few days.

The bad news with such bright cats who are clearly experts on how to shape human behaviours, is that they learn new ways of demanding attention without us realizing. The cat walking along the window-sill sees how quickly we leap to our feet to pick him up as he brushes past the family heirloom vase full of flowers, and simply heads to exactly the same spot next time he wants attention. Alternatively he may learn that chewing on a favourite plant or scratching the furniture is just as effective as threatening to spray. Like living with highly intelligent or hyper-active children, life with a bright cat can simply become a case of trying to keep pace with the latest trick.

Dear Mr Neville,

We've almost grown used to our Burmese cat, Baroness, spraying when she's upset or excited, or when we bring anything new into the house. We've always tried to be careful and give her plenty of opportunity and help to explore new things. If we have a bad period with her, we confine her to the kitchen for a while and then allow her into the rest of house again gradually, and this seems to work better than using drugs to keep her calmer. But last week she expanded her repertoire and included my fish and chips! They aren't supposed to spray food, are they?

Yours sincerely

Helen Drysdale

'No, Helen, they aren't supposed to spray food. Perhaps your fish and chips was of such poor quality that Baroness decided it needed a little extra vinegar before she could possibly consider eating any. I'd have a word with the fish and chip shop about the quality of their food which clearly isn't recognized as such by the cat, unless she was objecting to the quality of newspaper it was wrapped in.'

A flippant reply to cover my total disbelief that a cat should do such a thing and to cover up a feeling of helplessness about what to do. Baroness is recorded here simply to illustrate that yet again there are situations when cats defy explanation and treatment. We think we understand the mechanisms and motivations of some of their behaviour, and experience at dealing with problems leads to the development of ideas that can help with treatment. And then along comes a cat like Baroness and sprays what one has come to regard as a form of novel intrusion that cats will tolerate, and even a deterrent to spraying. It only serves to keep one on one's toes and never take anything for granted with a cat. As for Helen, she'll have to try the local Chinese take-away. Their offerings are usually irresistible as food, not spray posts, to all cats, oriental or not.

Middening

In case you are now tempted to put this book down to go and eat something, we will move on to another form of marking behaviour in cats, which is even more distasteful than spraying, that of middening.

Dear Mr Neville,

Since moving house a few weeks ago my cat, Snaggle, has persistently poohed both by the back door and in the hallway by the stairs. I've tried offering litter trays - in fact there are now six throughout the house - but if I put one where Snaggle has soiled he simply moves a little distance away. I have only caught him at it once, and when he saw me he behaved perfectly normally as if he had done nothing wrong at all. What do I do to stop this behaviour?

Yours sincerely

Anne McLeish

Middening is not a function of misplaced or misdirected toileting. It is a very definite act, using faeces placed in the open at specifically chosen sites, to identify the perpetrators occupancy or right to use a certain area or walkway. It seems most likely to be performed by pet cats which are already a little nervous or incompetent in their general demeanour, and which are then subjected to some form of major upheaval or challenge. The same types of upset that may cause one cat to spray may cause middening in another. Worse still, the behaviour is often more difficult to resolve than spraying because it represents a larger scale reaction to the challenge and indicates that the cat is more emotionally disturbed. The act of middening is usually carried out by doors or on walkways used by the family, as it is here that strange, unfamiliar or challenging smells from the competitive areas outdoors are brought right into the core of the cat's territory. But other cats choose even more public areas than this to midden. Some try to establish rights of passage on top of chairs, fridges and in their beds. Others midden in more personal items of ours such as stereo headphones. Electrical equipment such as video machines and televisions seem especially popular with the middening cat, and one acrobat I once saw even managed to midden on top of a door that it used to pause on mid-way between jumping from one cabinet to another!

Many species of wild cat, including tigers, do not always cover their faeces and seem to leave them in the open as a deliberate mark. Similar behaviour is observed in other animals such as otters and mink, which leave their faecal marks known as spraints, on rocks and high points along riverbanks to help define their territories. Middening is more often observed in feral cats living in high densities in our cities, than in farm cats with more freedom to move on if stressed or threatened by other cats. Dominant cats in any environment do seem to midden more often on high relevant points within their territory than subordinate or less assertive individuals, and un-neutered pet males are reported to midden in areas under dispute with other local house-cats. Middens are perhaps best interpreted as unavoidable blockbuster visual and odour messages, designed to drive the point home, as compared with the comparatively subtle and longer lasting, more complex patterns of odour left in squirts of sprayed urine. Pet cats which midden must be treated delicately along the same lines as sprayers, and not as dirty or house-soiling cats. Close confinement in a secure pen for a week or so is often the first step, with gradual exposure to as much of the rest of the house as the cat can feel confident in. Sometimes this will only be a couple of rooms but if the cat seems happy with that much I tend to advise that further progression is not forced upon it. Following identifiable disturbances such as Snaggle's move, controlled exposure to new rooms, neighbours and the garden is always to be advised and may be assisted by a little sedative treatment for a few days. It will also help if owners remember not to bring in too many items from outside for a few days, and to remove their shoes outside the door and put them in a cupboard before entering, until the cat gets used to the new range of smells in his new outdoor environment.

I get to see very few middening cats and this perhaps indicates how rare the behaviour is indoors though I suspect a great many of our pet cats, particularly in high density feline suburban populations, midden outdoors at disputed areas of territory or when arguments arise over which cat has access to certain points at what time. Such middening is most likely to occur away from the owner's property, often on a nearby neighbour's shed or garage roof, garden paths or feline walkways along the base of our garden fences. I also suspect that if cats start to midden indoors, owner tolerance and understanding levels are not likely to be high, and if the cat doesn't respond to the provision of extra litter trays (and I wouldn't expect it to, then it will not remain an indoor pet for long. If it is lucky it may get to be an outdoor cat with the same owner or it may go to a rescue home. But sadly, it may pay the ultimate price for being over-responsive to change or challenge.

10
Indoor Toileting

There is no animal quite so fastidious about its personal hygiene as the cat. Its meticulous attention to toileting and maintaining a clean home base is one of the main reasons behind the cat's popularity. While some dogs and even rabbits can be trained to use a litter box, no other domestic pet will dig a little hole to squat over, 'do the business' and then cover it up to avoid causing offence. Whether the cat uses the garden outdoors or a litter box inside, the vast majority of cats never make the mistake of leaving unwanted presents, except perhaps when ill, or when we accidently lock them in. The cat which is forced to toilet where it would otherwise not usually bears an extremely unhappy disposition for some time afterwards, so strong is the desire not to soil its home or sleeping area.

How should this behaviour arise and why is it that while young puppies may require some weeks of careful attention to become house-trained, kittens arrive at six or twelve weeks of age already abstemious in their habits? We can only guess that there is an evolutionary advantage, especially to the young, gained by not soiling near the resting area to avoid attracting the attention of predators or prey and to help ensure a food supply nearby. It is also suggested by American behaviourists that burying faeces helps disrupt the life cycle of some enteric worms and so reduces the risk of infection to the cat, and especially to vulnerable kittens. I am unconvinced by this suggestion because the tapeworm needs to cycle through a rodent's muscular tissues, and can only re-infect another cat if it eats the infected rodent. Perhaps a theory gone too far, though I have little to offer in its place except than that it seems to make sense that a cat, like a dog or a person, would be most unwilling to soil in or near its own bed because of the unpleasantness of lying close to waste, or in the damp. Those animals prepared to do so would also place themselves at greater risk of infection from bacteria, fungi etc. and so be at a disadvantage to those living in a dry clean bed. As a result they will be less likely to survive infancy to pass on their lack of hygiene to the next generation.

The cat's attention to hygiene in the nest is predetermined at birth. The young kitten is physically unable to urinate or defecate without stimulation of the abdomen and perineal (bottom) region by its mother's licking. This is known as the urogenital reflex and results in them only passing waste when the mother is present to clean them up immediately. Therefore the nest is kept clean and the release of odours which might attract predators when she is unable to protect them, is avoided. The reflex can remain operational until the kittens are about five weeks old though most can urinate and defecate voluntarily by three weeks. As anyone who has had to hand-rear kittens is aware, physical stimulation similar to the mother's must be carried out to ensure that the kitten excretes. This is usually carried out after feeding, though as the kittens grow up approaches can be made at any time and will cause the kitten to adopt a passive position on its back with legs open to receive the necessary attentions. Older kittens are more active and their own play and wanderings will serve to provide forms of stimulation and elicit toileting, or 'elimination', as behaviourists call the functions of urination and defecation. (This always sounds a little terminal to me.)

As the kittens start to wander and fall out of the nest to explore, their mother will continue to look after their toileting needs and they will demonstrate the urogenital reflex outside the nest. Indeed she may even carry them away from the nest for this purpose. This establishes not only the idea that the kittens are unable to toilet in the nest for hygiene reasons, but it also effectively trains them for life at the most impressionable of ages not to want to soil the nest. This early established antipathy helps ensure that subsequent generations will be taught similarly, and it is also an extremely useful feature to employ in the treatment of toileting problems when they do occur.

Outside the nest, all kittens have a natural tendency to paw and rake at loose earth or soft, movable material. Observing the mother's toileting behaviour and specifically her efforts to excavate a latrine hole and then cover it up after use, helps the kitten adopt similar toileting behaviour. Also the smell acts as an attracting focus when

133

nature calls and it directs the kittens into using particular latrines. By the time they are weaned at about six weeks old they are already using soft soil for their toileting functions. In short, kittens house-train themselves for the most part. All we have to do when allowing our cat to have kittens is to provide a suitable litter tray for the kittens to explore and learn to associate as their latrine in those first crucial weeks. The association with litter usually transfers well to soil if the cat is later allowed outdoors, so helping the cat perceive our home as his sleeping/feeding non-toileting area. For cats which are to remain indoors, toileting can be transferred to virtually any of the huge number of different litter products available for our cats while they are young, though most have their preferences and may be less willing to accept change later when adult.

The earth-raking behaviour of cats may occur in simple response to the presence of faeces and accounts for the enthusiastic covering by some cats long after the faeces have been obscured from view. They can still smell it so the raking behaviour persists. Others rake over the area nearby, or as is observed in many pets, the area alongside the tray and other objects within range even though they are different in texture and appearance to the litter substrate. For some the need to cover is very strong indeed and they will not only cover their own faeces, but also those of other cats. Observing cats at their toilet may not be the most rewarding or edifying of pastimes, but it is interesting to observe that the cat will often pause between rakes to investigate the smell of the covered faeces. Raking over continues until the cat is happy that the smell has reached the right level. This may mean total disguise, so as to avoid drawing attention to the depositer, or it may mean that the smell persists through the soil as a deliberate gesture to inform other cats of its presence. At this point, what appears to be toileting may have more of a middening function. Rather than toileting in the open, whereupon the faeces may weather very quickly, it may help to provide a fresher and more persistent mark. Again, our sense of smell lets us down. Even though the smell of cat faeces is highly volatile and detectable, we are unable to decipher that delicate package of odour information so as to learn exactly what is being said.

By and large cats are well suited to us because of the ease and lack of effort needed on our part to cater for their toileting habits. Either they are perfectly happy and conditioned to use a litter tray, and will strive to get there even when seriously ill, or they will use the garden. As our neighbour's gardens are further from the centre of our cat's world than our own, they may prefer to go there. However, this may occasionally cause some dispute, especially if your neighbour has recently tilled his vegetable patch and produced that soft top soil which is the favourite latrine of many cats. Others deliberately leave their faeces uncovered or only partially covered as a deliberate mark in their territory under challenge from other cats, and this too is perhaps more likely to occur away from their own home. But as our cat's territory of home and garden is also the challenged edge of another local cat's territory, we are probably just as likely to receive as our cat is to give, and between cat owners at least, all is usually about fair. Non-cat owners may be unconvinced about the beneficial acts of fertilization of their garden so freely provided by our cats, but if seriously disturbed, can always acquire a Jack Russell to ensure that no cat ever feels confident enough to adopt that most vulnerable of postures in their garden.

The upset experienced by the owner of a kitten which has never become so well house-trained, or perhaps worse, of the cat which has suffered a breakdown in previously immaculate toileting habits, can be severe. One of the fundamental reasons for choosing a cat as a pet is totally undermined when it leaves that unmistakable package in the middle of the carpet or behind the chair, or perhaps worse, a wet patch on the bed or morning newspaper. While most of us will accept an occasional accident, or a temporary loss of house-training due to illness, persisting problems are sufficiently disturbing to account for about one quarter of all the cases referred to me. This contrasts with the very small number of canine toileting problems I treat. Perhaps dogs are even cleaner than the fastidious cat, or is it just that we expect it more from a dog, or blame ourselves because we are ordinarily responsible for ensuring that it goes out. Perhaps dog faeces are simply less offensive, more openly deposited, or shorter lasting in their offensive odour than cat faeces.

Medical problems and their aftermath

Dear Mr Neville,

Since my cat Marcus suffered from cystitis some weeks ago, he has consistently urinated around the house. Solids are still well on target in the litter box but he now seems to urinate wherever he chooses. The cystitis has been cured by my vet so there is no medical reason for this to continue. Do you have any ideas as to why this should happen and is there anything I can do to house-train Marcus again?

Yours sincerely

Leonorah Jones

Medical illness, especially urinary cystitis and mild arthritis, which occurs most frequently in the older cat, are common causes for a breakdown in indoor toileting habits. While solids continue to be passed in the tray or outdoors, urine needs to be passed more frequently, so the cat may get 'caught short' more often. With general old age the cat's urinary sphincter may loosen, making retention of urine more difficult, and arthritis may also make retention more painful, and so the cat will need to urinate more frequently to relieve the discomfort. Such conditions may force the cat to urinate in places other than its usual litter tray simply because it can't get there in time or it is too painful to do so. Some cats understandably associate the burning pain of urination when suffering from cystitis with the litter tray and avoid it deliberately, selecting other 'less dangerous' sites. Once new areas are used as latrines, they may continue as such long after the medical condition and any pain has been relieved or cured. It is almost as though a new association to the different substrates of carpets or soft materials becomes established as equally acceptable to the cat as the litter. While many cases do resolve in time, most can be usefully assisted by providing the same association opportunities as the cat experienced as a kitten.

During treatment of the medical problems, it is often a good idea to house the cat in a much more restricted area than usual. One room, preferably a small one such as the bathroom, will ensure that he or she is never more than a few paces from the litter tray. If no small rooms are suitable then a greater number of trays should be provided. The aim is for the cat to travel as short a distance as possible should 'nature call' and maintain an association with the litter as his or her latrine throughout treatment, so precluding or reducing to a minimum the chances of urinating elsewhere. Each time the cat needs to urinate, he or she will be close to or on the litter tray, and so by a process known as successive approximation, will always get it in the right place through lack of opportunity to get it wrong. After a few days of consistent aim the cat can be allowed more freedom, with extra litter trays available in each room, until the medical aspects of the problem are resolved. The number of trays can be reduced steadily thereafter until the original number remains. However, older cats may always need closer confinement when unsupervised, or more trays around the house if particularly stiff and arthritic.

Similar latrine association problems have been encountered, though more rarely, with cats suffering from diarrhoea. This can occur as a result of gastric upset or as a secondary problem from upper respiratory tract infection, which may cause excessive mucus intake and lead to diarrhoea. Similar treatment can be applied with usually successful results.

Dear Mr Neville,

When my cat was ill my vet prescribed pills to be given three times daily. Fantasia, my cat, soon learnt to keep out of the way when the time came for her to have her pill and so I had to take advantage of her when she appeared. This was most successful when she was on her litter tray. We finished the pills and thanks to my vet, she is now recovered, except for the fact that she will no longer use her tray consistently for toileting. Instead she has started to use other areas, in the corner of the living room and behind a chair in my bed-room. I feel we are 'out of the frying pan into the fire'. What should I do to convince that there are no more pills to be given and to make her use the tray again?

Yours sincerely

Serena Cuthbertson

Oh dear! It's never a wise move to take advantage of anybody when they are at their most vulnerable, no matter how much easier it may make life at the time. Fantasia now clearly associates the litter tray as a dangerous place where unpleasant things may happen. The main thrust of treating this type of self-inflicted problem is for owners to make the cat's litter tray as secure as possible. In the same way as Fantasia is now selecting secure corners or areas behind chairs to give solid protection from interference, the owners should move the litter tray to just such a corner, or better still, in a corner and under a table to preclude 'attack' from above. The move itself will help reduce the cat's association of ideas. Still more important is to provide a covered litter tray by purchasing one of the proprietary brands, or simply inverting a cardboard box over the tray but inside the lip to ensure the litter stays in the box. A hole the size of the cat flap should be cut in one end for access, and once inside the cat should feel safe. This simulates the type of area selected for toileting by many cats outdoors, such as under a low bush next to a wall. Many cats are more confident and will toilet confidently in the open, especially on freshly raked soil, but like the proverbial ostrich, will turn its back to any onlookers as it squats and perhaps pretend that they are not there. Needless to say, to avoid such problems as Fantasia's, a cat should never be disturbed for any reason while toileting on a litter tray, and only outdoors if soiling where you are about to plant your vegetables.

Early learning

Dear Mr Neville,

I have recently taken on two three-week-old abandoned feral kittens and am having to hand-rear them. While I dutifully tickle their stomachs and rear ends after feeding, I have yet to notice any action on their part to dig into the litter that I place them on. Is it possible to teach them to dig, or will they do it in time anyway?

Yours sincerely

Linda Court

One of the problems of hand-rearing is that while we can now provide more than adequately for the kitten's nutritional needs, we still rely on instinctive behaviour patterns to develop without any help. Normally they do, and one would expect such young kittens to show the litter moving and digging reactions soon. However, we can never fully substitute for the mother's teaching, nor for the opportunity for the kittens to observe and learn from their mother's behaviour. With

138

this type of problem, it will help to manipulate the kitten's paws gently in the litter after their feeds and hopefully they will soon pick up the idea. It may establish itself more fully later when the kitten observes another cat digging a latrine hole. The important thing is to provide plenty of opportunity for the kittens to play in litter and to develop their interests in making it move, as a precursor to digging little holes.

Lack of early opportunity to experience soft litter or soil can result in the city feral kitten never associating it as his latrine. Even if successfully tamed later, some may experience difficulties in learning to use litter for toileting, having grown up on concrete or hard packed soil. Open toileting for urine and faeces becomes established and may be difficult to re-direct into what we would prefer the cat to use as a latrine. Fortunately, most cases can be resolved by the application of fine litters in a confined area for a while. Most seem unable to resist the opportunity to dig into fine, dry sand, or better still, one of the new generation of fine, light, particle litters such as Ever Clean from the United States, which has recently become available in the United Kingdom. Given no other alternatives for a few days, the behaviour soon switches.

Older cats given the luxury of a home later in life, perhaps having strayed away or been rejected from it earlier, may have been forced to toilet on solid surfaces during their spell of life on the wild side. They, too, may need a little time and forced opportunity to re-adapt to using litter again. With them it may help to provide soil in the tray initially but thereafter to change gradually to a preferred cleaner commercial litter. For feral cats which are adult when captured and tamed (ideally they should be neutered, treated and returned to site whenever possible), adapting to litter tray usage may be a very slow process in a few, but is usually surprisingly quick. Often the tamers erroneously interpret the use of the litter tray by the feral cat as the first step in its domestication. It isn't. The cat is simply doing what comes naturally.

A word of caution for all inexperienced cat breeders. Never be tempted to bring in soil from the garden as litter for young, unvaccinated kittens. Soil may carry many infectious agents which can cause serious illness or even death. However, the smell of soil can be applied to an ordinary cat litter and so ease the transition from litter to soil when the kitten has been fully vaccinated and can be safely allowed outdoor susually at twelve to fourteen weeks of age. The sterile smell of soil is available in a unique product called Tray Trainer by Vet Health, and I can vouch for its effectiveness because I carried out the necessary investigations for the company. It also helps attract cats back to using cat litter where a breakdown has occurred due to illness or behavioural problem and is generally preferred to untreated litter by cats both at home or in catteries. The smell of a cat's latrine is a key part in ensuring accurate usage. Kittens which miss the tray or have trouble in associating it as their latrine can be often be encouraged to use it by placing the 'mistake' in the tray rather than disposing of it. The kitten will be more likely to use the tray the next time and may even feel the need to rake over the previously misplaced effort.

Food proximity

Dear Mr Neville,

When we obtained our kitten we were delighted with how clean he was. We have lavished attention on him, bought him all manner of toys and established his own little corner in the kitchen for bed, food and litter tray. After a few weeks he stopped using his tray and began soiling the mat by the back door. He continues to sleep in his bed and eat his food, but will only use his tray if we put it on the mat, or away from his corner. We have only a small kitchen and this is rather inconvenient. Is there anything else we might do?

Yours sincerely

Sheila and Michael Pryce-Jones

Isn't it strange how many people assume that cats are so immaculate in their toileting habits that they would even be prepared to eat near their toilet. We certainly wouldn't dream of taking our dinner into the loo, yet in the urge to compartmentalize our pet's needs we may expect exactly this from the cleanest and most sensitive of our pets. In the same way as food can act as a deterrent to soiling and spraying in the home, it can also deter cats from using their preferred toilet, the litter tray, if it is placed too near. Food must always be placed as far away as possible from trays so that the cat is not forced to use other undesirable latrines. This is one of the most common reasons behind the onset of house-soiling by cats and definitely the most avoidable.

Latrine association

Dear Mr Neville,

My cat Fred developed quite normally as a kitten and was most clean in his toileting. However, at about four months of age he suddenly began to use the carpet in three or four places around the house and now refuses to use the litter tray for either solids or liquids. We have at least two trays in every room but he ignores the lot and simply heads to one of his places. If we move a tray there, he simply moves to a fresh spot. What should we do?

Yours sincerely

Jodi Burton

Just occasionally a cat may steadily decide that the litter provided or the tray and its position are not acceptable as a latrine area, even when food is nowhere near the tray. Sometimes this may arise because of more general nervousness or simply because the cat prefers to use the carpet. Once the new substrate is established it is very unlikely, if not impossible, that the cat would spontaneously revert to using litter again. Instead opportunity to soil the carpet is reduced so he will develop a preference for another substrate. This is achieved by confinement for a short period in a carpet free room, such as the bathroom or kitchen, or in a pen with a wooden floor. A litter tray is offered, lined with carpet. Usually the cat is impressed by this gesture, if not the confinement, and readily uses the box. This re-establishes the idea that toileting occurs in that place. The carpet is then speckled with a covering of lightweight litter such as Ever Clean or an alternative such as wood shavings or unprinted paper shreds. The layer is made thicker at every cleaning and after a few days of good, consistent use, the underlying carpet is entirely removed. The cat is then allowed progressively more freedom in the house and will usually stay bonded to the idea that toileting indoors occurs on a given surface, in a particular place and in a certain receptacle. For best results, the cat should have a period of at least two weeks before any contact with carpets is allowed without supervision.

Cleaning of unwanted toileting areas is as essential as cleaning up indoor spraying posts. The same cleaning system of Katastrophe by Vet Health, or the enzymatic washing liquid followed by a spray of alcohol as described on page 119 should be used on all soiled areas These should be allowed to dry thoroughly before the cat is allowed any access at all, supervised or not. Certainly no ammonia based cleaners should be used, and preferably none that contain chlorine such as household bleaches. Bleach, and the phenolic compounds found in many disinfectants, are very toxic to cats and their use must be avoided wherever the cat may roam, especially at his litter tray. The smell of a latrine is probably as important as the substrate and position of it, and serves to attract the cat and build up the association of that place as a toilet. Thorough cleaning to remove all traces of that smell is therefore essential, and unfortunately some deeply soiled carpets or furniture may have to be discarded to achieve this aim and for treatment to be successful.

Again, it is my experience that none of the standard proprietary repellents to indoor soiling are effective, nor are any of the traditional methods such as applying pepper dust to the soiled area. Food does however act as something of a deterrent and can usefully be placed at unwanted latrine sites once they have been cleaned (see page 122).

Dear Mr Neville,

For over eight months we have tried to get our beautiful Persian kitten, Khan, to be clean indoors but he just doesn't seem to realize what the litter tray is for. We have tried all types of tray, litter and places, but to no avail. He simply goes where and when he likes, even in front of our very eyes. He really has no idea of what to do, yet in other ways is normal and very affectionate. We have tried being kind and we have tried punishing him though we know it probably won't help. Please, can you help or we may have to put him to sleep, or keep him in a cage all of his life.

Yours sincerely

Anna Selwood

The desperation of trying to treat a kitten which just will not show any associative behaviour in its toileting activities can be enormously frustrating, as Khan is proving to his owners. For some reason, doubtless as a consequence of being highly bred, the Persian is the greatest exponent of all the breeds at failing to learn. Well over half of my house-soiling cases concern this breed, which is a pity, because they are often too expensive to carry such a burden. They are also too beautiful, if you like that sort of thing, to be dirty in the house and owners do indeed become exasperated in trying to get them to be clean. As Persians demand much attention with regards to grooming and maintenance, owners perhaps feel more cheated than most, and as few Persians are intended to live outdoors, the option of putting them there for most of the day cannot be considered.

With extremely serious cases of breakdowns in toileting behaviour, or with kittens, there is really only one course of action to follow. It isn't always successful, as the presence of other cats may be influencing the sufferer's toileting habits, but it has put a few die-hards on the road to indoor cleanliness. The cat is confined as before in a pen, but with only a bed and a small area of floor space. It is brought out to feed at meal times, and more frequently given access to water. The whole of the floor area is then covered with litter, preferably a fine, light brand. The cat will usually have developed the idea to move away from its bed to toilet, and so will do just that in the pen. If it does soil the bed, then this is removed and replaced with a cat size square of carpet. Then the cat's only option is to soil on the litter covered floor and gradually the size of this floor space is reduced. Progressively over about two weeks the area is brought down to the size of a litter tray. A plastic, side-less sheet is then placed under the litter, and then later a small-sided tray is used to house the litter. Usually the cat continues to aim for the litter throughout the period and gradually accepts a smaller and smaller area of litter surface as its latrine. The space where the litter used to be on the floor of the pen can also be occupied with tubs of dry food in the later stages of approximation to continue to force the cat onto the litter, again working on the premise that the cat will not voluntarily soil on or near its food. The only drawback with all this is that Persian cats will have to be cleaned and groomed far more frequently than usual, especially in the early stages of confinement. But if all goes well, the attachment to litter is usually as strong as with any other self-taught cat, and the patient can be allowed out of the pen into one room and then gradually into others.

Why Persians? There may be some inbred inability to show interest in loose litter early on, or perhaps they are simply slower or unable to associate their mother's latrine behaviour as being something to emulate. More likely is that the behaviour is undeveloped in the mother and the kitten has no opportunity to learn by observation. Poor mothering ability in any cat, pedigree or moggy, can cause all types of problems in kittens from inappropriate toileting behaviour right through to death from neglect. But poor mothering ability is not the reason behind every case, as perfectly house-trained Persian mothers may also produce an occasional kitten which doesn't follow her good example. So really we need special efforts to be made by breeders of Persian cats to identify such kittens as early as possible, so that they may be given special attention, with paw-guiding for digging and investigatory play in litter. Perhaps then I'll see a few less, though to be fair to Persians generally, I don't see that many. It's just that they always make the most difficult patients when it comes to house-training.

Ann Selwood had been moved to punish Khan for his toileting errors indoors, as an act of desperation and much against her better judgement. All punishment is pointless in trying to treat the indoor toileter, especially the old nonsense about rubbing the cat's nose in it. The poor cat will have no idea what is happening after the event other than the fact that its normally friendly, mother-like owners are behaving in a frightening and unpredictable manner. The same will be true even for the cat 'caught in the act', for whom there is no crime other than wanting to relieve him or herself. Punishment will not help redirect the toileting behaviour into the waiting litter tray, or make the cat anything other than nervous about the approach of its owners. And a nervous cat has less control over its urinary and anal sphincters. Patient application of a few simple techniques is the only way to treat the cat that soils indoors, and never the use of negative conditioning.

Cleaning

Dear Mr Neville,

When we only had Morris there was never a wet patch in the house, other than in the litter tray. Then we decided to get Morris a little friend, Marla, and ever since there have been occasional wet patches made on the kitchen carpet near the tray. Should we clean the tray more frequently now 'we are two' or provide extra trays?

Yours sincerely

Brigette and Olivia Simons (age 9 and 11)

Cats do not like to use dirty litter trays and doubling the input by doubling the number of cats is a sure sign to double the number of trays, or better, to double the frequency of cleaning. The main reluctance probably stems from the cat's dislike of standing on damp surfaces and smell of litter that has been over used. It may help to change the litter gradually to a more absorbent variety which clumps when wet and so concentrates the smell and damp. However, there is little substitute for regular cleaning. Some cats do not like to share trays even with friends and there is little option here other than to provide more trays or clean out existing ones after every usage. Again the clumping litters may mean less work for the reluctant cleaner as only the clumps and faeces need be removed for a while, and complete changes of litter will be required less frequently.

Substrate preference

Dear Mr Neville,

We used to use a grey earth type of cat litter for our cat, but found this very awkward to carry up the stairs at our new flat. It was even heavier coming down once used so we changed to using a scented wood pellet variety. Since then, Tom has been most reluctant to use his tray and we have now had to go back to using the grey earth. I would have thought the pellets would have been more hygienic for him and would very much like him to use them. Is there anything we can do?

Yours sincerely

Pauline and Bertie O'Sullivan

Cats can bond very specifically to one type of litter and a sudden switch from grey earth granules to wooden chip pellets, with added pine or other scent, is a far greater change than we realize for the cat. Many are unimpressed by the advantages of such a modern lightweight approach and others are seriously deterred by the scents which are added to some litters to please the human purchaser. This seems especially true of those which release the bulk of their scent as they become wet from the cat's urinate. The whiff of fragrant pine may deter the cat from raking, and perhaps from using the tray altogether. Discouraged and not a little confused, the cat will often appear desperate to seek other less disturbing places than his tray, so quickly building up an association with the carpet or elsewhere as his latrine. So, choose your cat's litter with care, and never introduce changes suddenly, as these can cause a breakdown in toileting habits very quickly. I suspect that because of their light weight, long tray life and attractiveness to cats, the fine grain litters such as Ever Clean will soon become the most widely used litters in the United Kingdom as they have in the United States. The days of hauling great sacks of heavy earth litters may well be numbered.

Where a change is to be made, the old litter should be phased out gradually in favour of the new over the course of about a week, so that the cat slowly acquires the new association of texture and smell. A small amount of used damp litter from the previous day should be added to the new mixture to encourage smell association. This technique can also help cats start using a tray again where there has been some breakdown, as it helps them recognize the latrine for what it is. The trays of cats with house-soiling problems should never be cleaned out too often as this precludes the development of the latrine association. Once every two days per cat, that is once a day for two cats, is about right for the average size tray and this can be increased to once per day per cat once the problems of soiling elsewhere are resolved. This is a little less hygienic than could ordinarily be recommended, but providing the usual precautions of washing the trays and wearing rubber gloves are observed, the risk to the owner is not appreciably greater, and the treatment will help. Of course the cat must be up to date with anti-worming treatment at all times.

Dear Mr Neville,

Our cat Roger is not the brightest of cats but gets through life with cowardly bluff. That is, until he misses his tray. He knows enough to go to the tray but seems unable to aim things in the right direction. Most times he climbs in the tray and then hangs his bottom over the side to deposit his offerings on the floor. If we put paper under the tray, he doesn't get in the tray at all and simply squats by the side of it. Is Roger stupid or are we doing something wrong? Please help.

Yours sincerely

Mary and Alan White

Roger has at least got some things right. He clearly knows that the tray area is his latrine, but not how to position himself, or how to deposit urine or faeces into the litter itself. It seems that Roger doesn't dig a hole either, so perhaps his early learning was incomplete or was interrupted at a crucial stage. While some cats can simply be treated by offering a larger tray, or one with higher sides, the usual form of treatment is once again to remove the opportunity for error by confinement, and by covering the tray to ensure that he cannot hang his bottom off target. Many cats whose attachment to litter trays is incomplete, are reluctant to enter the confines of a covered litter tray as they do not see it as the secure area that most cats do. For them it is perhaps a restriction that builds on an established antipathy towards using the tray. Therefore a roofless cover helps direct the aim without the cat feeling restricted and, if coupled with a preferred finer grain litter, Roger and his affable type shouldn't continue to make mistakes. If possible, such cats should also be shown how to dig by manipulating their paws if they allow it. Where such an effort frightens it, the movement should be discontinued immediately so as to prevent the cat from developing a reluctance to go anywhere near the tray.

Dear Mr Neville,

Puss, our tabby female cat, will often urinate in the bath or kitchen sink rather than in her tray or outdoors. Is there any reason for this? It isn't a great problem as washing it away is easy, though we do have to remember to do so before

147

washing up or having a bath.

Yours sincerely

Stella and Norman Smith

This is a habit of quite a large number of cats, but is usually only problematic if solids are deposited. There may some latrine type smells emanating from the outside drain which percolate up through the plug to attract the cat, or it may simply be that the sink or bath is one of those nice secure toilets protected on all sides and therefore more attractive than open trays. It is always worthwhile looking closely at how the cat uses existing facilities and making them more attractive by moving them, covering them or changing the litter. Thereafter, simply filling the bath or sink with a few inches of water when not in use will deter most cats, though if this can be coupled with offering a normal tray in the same room, the cat will hopefully divert to it when finding the bath or sink latrine unacceptable. One can go as far as sprinkling litter in the bath and then in a litter tray in the bath. The tray is steadily raised up and out of the bath over a period of a few days, moved over onto a chair and finally onto the floor and out of the bathroom. This seems a bit long winded but will usually work. It can't be recommended in the kitchen for hygiene reasons, so the water technique is better all round.

Nervous urination

Dear Mr Neville,

Mimi is such a timid cat. She seems frightened of every loud noise or sudden movement and runs away immediately when visitors arrive. While she normally is very clean, after any disturbance she may urinate in a corner or under the bed or sideboard. On one occasion when a relation brought his dog when visiting, she even defecated under the bed in fear. Is there anything we can do to make her feel better and our flat cleaner?

Yours sincerely

Miriam and Charles Bewley

Nervous urination is one of the most common forms of indoor soiling that I see, but nervous defecation is fortunately a rare entry in the diary. I well remember the rage of one owner whose cat had managed to leave a smelly pile in his hi-fi headphones, but as the cat was rehomed post haste, I was unable to effect treatment. If the owner's temper and lack of humour was indicative of a general intolerance, then I suspect that treatment of what was a very nervous cat would have proved most difficult anyway.

Treatment of nervous urinators and defecators involves the application of the suggestions in chapter 5 in order to raise the cat's competence at coping with challenges, such as the arrival of visitors. In severe cases this will involve controlled exposure to such challenges by penning the cat, and so a litter tray should be provided to force the cat to use it in these situations. Raising the animal's competence is the main thrust of treatment but support drug therapy with progestins is not recommended for nervous toileters. Progestins can cause water retention as a side effect and make the cat need to urinate more suddenly and at higher volume. This would be counter-productive to treatment and so other forms of mild sedation such as valium, or some homoeopathic remedies, should be considered first in such cases.

The cleaning of areas soiled by the nervous urinator is essential to prevent them from becoming established as latrines, but as most nervous eliminators urinate or defecate involuntarily, there will usually be few sites used more than once.

Some nervous urinators only do so when chastised by their owners, and by contrast, these cats may head for a sheltered area to hide and then urinate or defecate. Never was the point so well made about being consistent with a cat and avoiding punishment for misdemeanours. Owners must always be pleasant because for some sensitive individuals their whole security base collapses when their owner directs aggression or threats at them. Most cats are not quite so sensitive of course, and may need an occasional reprimand when attempting to climb the curtains or walk across the table during Sunday lunch. The important thing is to make such intervention more startling than punishing, and as untraceable to the owner as possible. A loud hiss by the owner will be interpreted as a general warning/threat in feline terms and inform the 'offender' to escape from the area he is in. However, use such methods with caution with the nervous cat, especially those which urinate when disturbed.

Associative elimination

Dear Mr Neville,

Last weekend my wife and I went away for a break and left our cats, Tom and Jerry, the run of the house rather than board them. Our neighbour came in twice a day to feed them, clean their trays and see that all was well. They continued to eat well and seemed undisturbed by our absence with a very unpleasant exception. They used our bed, favourite armchairs and a few clothes we left on the bedroom floor as toilets. Solids and liquids awaited our neighbour every morning and naturally she is now unwilling to look after our cats when we go away in future. What should prompt a cat to do such a foul thing, and more importantly, how should we prevent it in future?

Yours sincerely

Gwen and Harry Clarke

This type of indoor soiling is sometimes classed as 'resentment' toileting but probably rather unfairly as although it seems a deliberate act of vengeance or revenge, it is probably more the action of a disturbed or very nervous cat. Some go as far as urinating on the owner's lap, and a few more will occasionally urinate or defecate on the bed while the owners are in it. It is very hard to feel sorry for such cats at such times but avoiding punishment is essential. The suggestion offered by the editor of this book that owners should take revenge on such cats and pee in their bed in return is also not recommended! Short term fears or frustrations such as the withdrawal of a protecting influence when leaving the home for a few days, or even the arrival of a new baby can cause the onset of such behaviour. However, it is essential to review the position and acceptability of the litter trays or access to the outdoors for toileting and to ensure that it is adequate for the cat. Assuming that the cat has perfectly reasonable facilities and is not simply wandering in search of a toilet, we can usually assume that urinating on piles of dirty clothes, or the bed or a favourite armchairs occurs when the owner is absent and it can be interpreted in one of two ways. The first is that the anxious cat is simply sleeping or resting where it normally sleeps, but on waking feels nervous or isolated and involuntarily urinates or defecates. Having done so once, the area is then seen as a latrine rather than a bed and continues to be used as such, with other more sheltered areas being sought as resting areas. The second and more likely explanation is that the insecure cat actually seeks areas and items smelling most strongly of the owners and deliberately urinates or defecates on them in an effort to associate his or her smell with the supportive smell of the absent owner. This is more an act of marking behaviour and is designed to present a stronger front to any enemies, real or imagined, that may see the owner's absence as an opportunity to challenge the cat's occupancy of the home. When such elimination occurs on the owner's lap or the bed while they are in it, we can guess that the cat is emotionally upset for some reason and having found protection on it's mother figure either tries to endorse the bond or simply over-relaxes and involuntarily urinates in relief.

This may be very interesting in theory, but if nothing else it demonstrates the need to isolate cause and effect. If the cat is simply trying to keep itself confident by an association of smells because it feels vulnerable in its own home when the owners are absent, then it may be better and kinder to board them in a secure cattery, (run to the standards suggested by the Feline Advisory Bureau), rather than leave them at home alone. Owners sometimes have a stronger influence on the security of the cat than perhaps we realize, yet it would seem integral to the substitute mother idea. It may be possible to provide greater security by leaving the cats in a smaller area in the home which should not include owners' bedrooms or the living room, so as to preclude the possibility of associative urination. Nervous or associative toileting should then no longer occur as the cat will be secure in the smaller predictable area. Such techniques are also employed to help the dog which cannot cope with life alone and may suffer from similar separation anxieties. Dogs normally howl or destroy household items in response to anxiety, but may also lose toilet control. It is never as directed as the cat's loss of toilet control in such circumstances, and so is seen as far less of a considered action to maintain security.

If boarding such cats is not acceptable, then lodging them with friends or employing a full time cat sitter may be the only alternative, but practising short separations and improving overall competence at coping with novelty and challenge are vital for the permanently indoor cat. In serious cases, the use of progestin treatment from the vet can help, though only for the short term to avoid the problems of water retention. Such prescription should ideally start two or three days before the intended departure so that the cat is prepared physiologically for the shock of the owner's sudden absence from home. I would again suspect that many alternative treatments could have a role to play in treating such cases but as yet have not had the opportunity to test their effectiveness.

One unfortunate after-effect of this type of elimination problem is that having started to soil on beds, especially on duvets, it is very difficult to break this habit. Soft materials continue to attract, and cats must be denied access to beds for several weeks to break the association - that is, if they haven't already been barred for life anyway. This should be accompanied with a litter more suited to the cat's toileting needs and the owner's expectations.

So much for the messy end of the job, though of course it does account for much of my caseload. Now we shall turn to the far less understandable or explicable aspects of feline behaviour, and for which there is often very little on which to base treatment other than sheer guesswork. Sometimes it works, sometimes the problems go away of their own accord and sometimes they don't. But whether treatment has been effective or not, the final chapter of this book is dedicated to those cats who demonstrate bizarre or totally pointless behaviour, and also to their sometimes long-suffering owners.

11
Strange Diets
and
Bizarre Behaviour

Many of the problems that occur with cats are a result of their excessive attachment to us and inability to function competently without us. They also bond strongly to the territory we provide around our shared home base. Moving home can therefore be traumatic for the cat as it involves a complete change of the whole territory, destroys walkways and escape routes and can leave the cat feeling totally vulnerable. Slow careful introductions, one room at a time, and lots of attention will help most cats over the stress of such upheaval within a couple of days and few will need any sedative support. It may be wise to board particularly nervous individuals in a safe and secure cattery before the stripping and packing starts at the old house, and not to bring them to the new house until everything is unpacked and positioned. Outdoor cats with a wider experience of change generally cope better, but should be kept in the new home for a week or so to learn the geography and smells of their new base. When finally let out to explore their new outdoor patch and carve out a piece for themselves with local resident cats, they should be hungry. Starved of food for twelve hours or so, they will not wander too far from home and readily respond to the call or plate bashing that signifies 'dinner is served'. Accompanying the cat on its first few excursions into the brave new world will also help, but the cat's adaptability and survival instincts usually serve it well, and they soon adopt a similar lifestyle and habits as at the old house. There are exceptions however.

153

Moving home

Dear Mr Neville,

My partner and I moved into a new flat about six weeks ago along with our ginger cat Banzai. We kept him in for a couple of weeks and he seemed to settle in well. However, since he has been allowed outside again, he keeps returning to our old house which is about two miles away. Is there anything we can do to tell him that we, and he, have moved?

Yours sincerely

Madeleine Fosket

This is a very common problem. By moving so close to the old flat, it is highly likely that in exploring his new territory Banzai will have encountered old routes through his previous one. He will simply return home along those routes as before and then look confused on arrival to find that all has changed. The bond with the new home is simply not well enough established to attract him to it. Some are inadvertently encouraged by the new occupiers of the old base who provide food, or who are flattered by this strange cat's confident entrance through the cat flap and willingness to set up home with them. But even when they have been warned that the cat might return and take repelling action by turfing him out or throwing water at him, the bond with the old centre of the territory can persist. The cat will keep returning and will only go to the new home if physically taken there. Both parties can get tired of the travelling, especially in the remarkable cases where cats have returned to old haunts many miles away. Where cats have travelled thirty miles (I have encountered two of these hikers) or where epic journeys are made across the country covering hundreds of miles, there can be no possibility of this being due to encountering old routes. Cats, like homing pigeons, migratory birds, salmon, whales and even some dogs must have an inbuilt homing sense that they use to navigate their way back. Most theories suggest this sense is a magnetic one and that the cat has an inbuilt compass, responsive to the earth's magnetic field. However, others suggest that it is more a finely tuned chemical detection system that enables a cat to walk up a concentration gradient of scent towards its home where the familiar smell, or combination of smells is strongest. Treatment of these homing specialists may be impossible with our present lack of knowledge and therefore including the cat in the sale of the first house may be the only course to take. Some have suggested attaching a magnet to the cats head to scramble the homing sense but I have no information to suggest that this would work. One to try for the future perhaps.

Cats such as Banzai can usually be successfully treated. The first step is to ensure that the new occupiers of the old house do everything to detach him from his old home by chasing him away and throwing water at him, and never stopping to say hello or feel sorry for him. Other neighbours, even if previously friendly must be asked to behave similarly. Banzai should be kept indoors at the new house for about a month, but if he continued to return, should never be taken back to the new home by a direct route. Instead the owners should make as wide a detour as possible, heading off in the opposite direction from the new home and driving a good few miles before circling round. Sometimes it even helps to board the cat for a few weeks in a cattery as far away as possible from either home in an attempt to scramble both memory of the old home and any homing mechanism. But once at the new house, again the tricks of short frequent feeds and plenty of love and attention against a secure background should help build new bonds. The new home should come to be perceived as the centre of the new territory and a source of food and shelter, in contrast to the unpleasantness of the old home. It may take weeks, and in some cases months, before the cat can be allowed outside unattended. The moral of the tale is always move at least five miles to a new home.

Where some cats bond to a territory so emphatically that they find it hard to change, others can become so attached to the owners that they find it hard to cope without them. Over-dependence can lead in old age to increased attempts by the cat to maintain contact vocally. This lack of security is understandable since it regards its owners as protecting mother figures. The pack dog frequently suffers from separation anxieties when isolated from us, its substitute pack, and can make desperate efforts to keep with us. Trying to head us off at the door, guarding the exit and the cupboard where the coats are kept, are all designed to maintain personal security by staying in a group. And if departure is successful, the lonely, anxious dog may howl and cry to attract us back or try to escape by digging at doors and climbing out of windows in search of us. When complete panic sets in the dog may lose toilet control or become destructive of household property.

This is not observed in cats, yet many are clearly unhappy and insecure when separated from their owners. Some do cry a pitiful high wail or short mew, similar to the distress call of kittens. Others make concerted efforts to stay with us, the most remarkable case being that of a Burmese cat, a four-year-old spayed female and one of four in the house, which could not face life alone at night without its owner. This cat scaled the outside of the house via the guttering and ledges, to the eaves of the roof outside the owner's bedroom, three storeys up. It crawled in through a hole and then wailed to attract attention. This happened every night for about three weeks. The owner naturally responded to the awful din by getting up in her nightdress to rescue her via the attic and then keep her with her in a nice warm bed all night. This is a typical case of learned incompetence, now thankfully resolved by a little social detachment between owner and cat, a restructuring of their relationship so that affection is only initiated by the owner and never the cat, and mainly by offering the cat an extremely attractive heated bed downstairs in the kitchen. The case served to indicate how strong the bond can be between solitary predator and owner, when we provide warmth and protection.

Dear Mr Neville,

Occasionally my wife and I have to go abroad and because of the quarantine regulations, are forced to leave our beloved cat, Shona, behind. Usually we take her to a cattery of impeccable hygiene and comfort, but find that she is most distressed when left. She will barely eat, spends much time in her bed and if pressed into a little activity by the cattery staff, may even lash out at them. Yet as soon as we arrive back to collect her, she comes running like a dog to see us and even seems to recognize the note of our car engine. She purrs furiously and immediately settles back into our arms. Once home again, she eats furiously and follows us around the house as she normally does, apparently none the worse for her 'holiday'. Can we do any more to help her bear the separation better?

Yours sincerely

George and Julietta Barclay

Whether cats like Shona are highly attached to their owners or simply dependent on the whole nature of their lifestyle, is sometimes difficult to assess. Most owners would rather believe that it is they for whom their cat pines when they leave him or her in a cattery. This feeling that the cat is distressed by their departure compounds the burden of guilt which owners bear as a result of being unable to take the cat with them on holiday. No matter how much time and effort cattery owners invest in making their establishments clean and attractive, owners may always tend to regard them as temporary prisons. The strength of the bond between the cat and the owners can be at the basis of the cat's security to such an extent that separation anxieties result when the cat is taken away to even the most sumptuous of catteries. Some boarding kennels are now trying to offer the creature comforts of home for boarding dogs with chairs and beds to climb on, televisions to watch and high staff-to-dog ratios to keep them amused. Perhaps this will start to happen soon with cats, though I doubt it will bring much improvement for cats like Shona whose home, routine and owners are such a vital feature of their well-being that all else is inadequate and frightening. Treatment of such occasional anxieties is very difficult but some progress can be made by preventing the cat from following its owners so much, when it is at home and being generally less responsive to the cat's demands for attention. More important still is for the cat to go to the cattery for a few hours on a few occasions before the real boarding period, so as to get to know the cattery staff while the owners are present. All comforts from home such as beds, favourite toys and old clothes smelling of the owner can also help the cat transfer its loyalties, though this is likely to have limited success, and will probably only be possible with one particular member of the staff. The most acceptable person should also be encouraged to visit the cat in its own home to make friends and take over the role of feeding and fussing. Meanwhile the owners become more aloof. The aim is for the cat to develop the ability to cope with the change of home and lifestyle and to be assisted by the new temporary carer. It may only be possible in a very few catteries with very special staff, and far more cats simply just have to suffer from their separation anxiety. Whenever possible, such cats will usually be happier if left at home to be cared for by a cat/house sitter or 'pet aunt' who will call to feed and care for the cat. At least the cat's whole world is not disrupted and it can continue to be surrounded by familiar territory and comforts, if not its owners or familiar routine.

On returning to a cat left behind, the owners will often be greeted by a huge display of feline greeting with little yells, purring and rubbing as the cat rediscovers its source of maternal security. We respond to a cat's display of affection or flirtatious greeting postures by issuing a stream of high notes in the same way as mother cats and their kittens communicate with each other. This can occasionally have amusing repercussions.

Musical cats

Dear Mr Neville,

My cat Mignon loves everyone. She rolls on her back invitingly and adores being stroked and fussed by any visitor, or even people she meets in the street. She is always purring and is quite the happiest cat I have ever known. If I whistle she comes like a dog and starts to roll, purring loudly, but if I whistle high notes, she leaps to her feet and runs up my body to rub around my mouth. She grips hard with her paws around my neck and rubs continuously with the sides of her mouth against my face and under my chin. Sometimes it's easier not to whistle at all, but as a musician I simply must practise my flute playing. This sends Mignon into a rapturous frenzy as she leaps onto me, digging her claws into my leg and rubbing furiously around my face and the flute. It's quite impossible to practice, but if I shut her out, she claws at the carpet by the door, desperate to get in with me again. Is she just a musical cat or are there logical explanations for this strange behaviour?

Yours sincerely

Jennifer Hillborough

The high notes of our whistle, or even better the purer, higher ones of the flute or other lighter wind instruments, seem to mimic the sounds made from a kitten to its mother. Having attracted the mother, the calls cause her to rub the kittens and physically reassure them by licking. This often happens around the face of the kitten. Mignon is probably seeking face to face contact with her flute-playing owner because the sounds, no matter how well practised, indicate distress, and she wants to offer her owner reassurance of the maternal kind. Role reversal in our usual relations, and probably unlikely in most cats. One would suspect that queens which have had litters of their own, or particularly friendly cats used to much vocal communication, would be most likely to respond in this way.

We all talk to our cats in high pitched tones because they seem more responsive to them than to our normal voice or low pitch noises. Quite apart from the felinese gobbledegook that we spout to attract their attention, we also use very high squeaky 'come hither' sounds, and the cat responds once again as if it were a kitten responding to its mother's call. They trot towards us with tails high, expecting to be greeted and petted by a friend, and we keep up this form of greeting throughout their lives. The combination of sounds is almost impossible for us to detect, but as cats will usually not respond as readily to strangers making similar noises and offering similar affection, we can presume that each of us has our own personal 'squeak', recognized by our own cat and few others.

The hearing ability of the cat is approximately the same as ours at low frequency but far superior at high frequencies. We can hear little above 20 kHz and a dog, despite a reputation for good hearing, detects little above 30 kHz. But the cat responds to notes of 65 kHz, at least one and a half octaves beyond our limit. So our high squeaks are well within their range and their general ability to detect such high notes may account for the jumpiness of some cats around our household electrical appliances, such as the washing machine.

However, for super-responders like Mignon there may be no alternative but to banish her outdoors or two rooms away during flute practice; or else there should be a little less tolerance of her efforts to reassure the owner when she is emiting those distress calls.

Again we tend to suffer more from dogs which demand attention than cats. The demanding dog can employ a variety of ruses to gain centre stage, from standing in front of us wagging its tail to leaping up at us, barking and carrying off slippers, wallets and handbags as trophies. Cats are usually more subtle and calmer with their demands, though a few will run up legs to get to the cuddling zone. Fine if you're wearing jeans, painful if you're in a skirt and tights. Most cats will seize the opportunity for a little maternal comfort and affection by jumping on our laps whenever we sit down, recognizing that we are then available, relaxed and most likely to be responsive to demands for attention. Failure to respond on our part usually causes the cat to move away in a state of obvious 'huff', his back turned and head held high and aloof. A few go a step further and insist that we provide some form of attention by protest spraying (see page 127), and a few others will engage in other activities designed to make us scurry towards them. A favourite trick by these cats is to leap onto window sills and tread less than gently through our favourite pot plants or most prized ornaments and antiques. It works every time. Thankfully most demanding cats stick to repeated efforts to climb on our laps, or simply paw at us, and when met with no response, simply pretend that they don't care anyway.

The nature of our relationship with the cat has been referred to as maternal throughout this book and never is this more apparent than with the many cats who, even in the prime of adult life, will readily suckle on our clothes and occasionally even our skin. They salivate profusely during such behaviour in anticipation of the milk feed that it used to bring them as kittens while their paws and claws make a stepping, contact and relax motion which used to be directed at the mother's nipples to stimulate milk flow. With a big bruising adult cat this behaviour can seem incongruous with his outdoor despotism towards all other life forms and not a little painful and damp for the owner victim. This behaviour is seen in nearly all pet cats to some degree and is triggered by our physical stroking and handling, or simply by our reassuring presence. While this may be tolerable, the cat which dribbles all over its owner is usually only allowed near for short periods and turfed off when the saliva starts to flow. Kittens which have been hand-reared may demonstrate this behaviour more readily and with greater gusto for the obvious reason of full maternal substitution.

Some owners, particularly with rather nervous cats or those rescued from an unfortunate beginning, may be unwilling to reject the cat for fear of contributing to further rejection or withdrawal of care. Undiscouraged, the cat may continue with this behaviour to the point of bringing their owners out in rashes, or rendering corners of their clothes wringing wet and well chewed. While it is easy to suggest that such cats be discouraged from perpetuating infantile behaviour patterns, there may be a strong reluctance on the part of the owner to reject the cat by putting it down onto the floor or by being less available for any affection. Keeping encounters short and frequent with adult interactions of rubbing and stroking do soon replace the unwanted infant patterns, if the owners so desire. Others, particularly those more dependent than most on their cat for company, such as many elderly or housebound people, actually prefer the behaviour to persist and are prepared to bear the damp consequences at the price of having something warm and responsive to care for.

Such cats seem especially interested in sucking woollen items of clothing. This may be because the texture of wet wool may be similar to that of a mother cat's fur around the nipples, or more likely because the smell of lanolin and other natural odours in wool, are similar to those in the nursing cat's fur. But wool sucking is a different behaviour to the wool eating that I have encountered in many oriental cats. Wool and fabric eating, or the eating of any non-nutritional item is termed 'pica', which strangely is the Swahili word for cat.

Pica – fabric and wool eating

Why some cats should want or need to eat wool, and indeed other fabrics, is not understood. That they do is beyond any doubt as many owners of clothes, carpets and furniture covers with holes in them, can testify. In the 1950s the behaviour used to be thought of as a trait restricted to certain strains of the Siamese breed, and particularly of queens in oestrus. As it seemed to be an inherited disposition, some attempts were made to breed it out of these strains. The results of a survey that I have been carrying out since 1988 reveal that the problem is more widespread and not restricted to the Siamese. Of 152 cases reported at the time of writing, 59% are of the various types of Siamese, 28% are Burmese with 13% of mixed parentage. Some of this latter group have one parent Siamese or Burmese, but some do not, and are apparently pure 'moggy'. While their exact family history is unknown and may contain some orientals, it is likely that the condition simply occurs less frequently in crossbred cats. It is perhaps more logical to think in terms of some genetic abnormality caused by malfunction at the chromosome level, which is more likely to occur in cat types which have been closely bred. Of those responding to the survey who had contact with owners of their cat's littermates, 58%

said that at least one of their cat's brothers and sisters was also a fabric eater. Not conclusive, but it does suggest that the condition is inherited to some extent at least.

The most likely explanation for fabric-eating probably lies in a malfunction between the brain and neural control of the digestive system, though the degree of involvement of the central nervous system ('voluntary' decision making) and the autonomic nervous system (involuntary nervous action such as control of digestion) is as yet unclear. Also the relative frequency of this behaviour as shown in the survey results, is probably more a reflection of the willingness of certain owners to reply to the questionnaire than as a measure of true occurrence.

Dear Mr Neville,

My Siamese cat eats not only wool, which we keep well out of her way, but also all other fabric. She has a particular passion for towels and tea-towels, socks and T-shirts and has now started chewing at the settee cover. She will chew in front of me, and even on me if she gets hold of my jumper, and seems unperturbed by our screams of disapproval and even squirting water at her. If we go out I now lock her in the kitchen where she can't get at anything. Is there anything we can do to stop this expensive behaviour?

Yours sincerely

Terrie Peters

Some wool/fabric eaters do stick to consuming wool, and again one might expect that the smell or texture of wool is the initial trigger to the behaviour. But the majority broaden their appetite and will consume all fabric. Items of clothing, preferably worn, and towels are especially popular. In the survey, 93% consume wool, 65% will eat cotton and 54% synthetic fibre. Perhaps the scent of our clothes attracts the cat and stimulates the need to consume them. Why two cats in the country at least should restrict their fibre eating to the discarded knickers of the lady of the house is unclear, and too rude to investigate further. In one of those cases, the last ritual at night before retiring is to locate and remove the cat from the bedroom where it has cunningly concealed itself in the hope of a successful abduction and consumption of the discarded Janet Reger's.

When eating any fabric, be it underwear or less personal items, the cat will appear totally engrossed in its activities and will sometimes be in a trance-like state. Intervention by hissing or yelling or even throwing water at the cat may cause him to stop but often they will simply go straight back to the item or look for another in a quieter place. The cat will take in the wool with its canine and incisor teeth and having obtained a good mouthful, will start to grind it up using its shearing molars at the back of the mouth. The volume of fabric consumed by some cats is truly remarkable, and even more so when one considers that it usually passes through the cat unaltered without causing any harm. Unfortunately some do suffer blockages in the stomach or further down the digestive system and a few are euthanased because of the resulting damage, but some live long healthy lives, eating wool or other fabric every day without any repercussions.

A few more are destroyed because their owners find their habits too expensive to live with. Some have damaged dresses worth hundreds of pounds or furnishings worth thousands, and the average is £136 per fabric eater. Interestingly most who replied said they had learned to live with the problem and would not dream of parting with the offender, nor would they be deterred from having a cat of the same breed in the future.

One theory as to the origins of fabric eating suggests that, like wool sucking, the behaviour is linked to a continuing infantile disposition in what are traditionally sensitive breeds. Thirty-nine per cent of cats in the survey exhibited the usual forms of continuing infantile patterns such as excessive kneading of their owners, suckling and salivation when petted and perhaps an over-attachment expressed by the need to stay in physical contact with them. The age of weaning of kittens may also be crucial. Kittens taken from their mothers at six to eight weeks of age may be more likely to show stressful and nervous responses later in life than those taken at twelve weeks. The extra weeks, though perhaps unnecessary from the nutritional point of view, may be essential for the emotional development, so that it can later cope with challenge more effectively and so be less likely to reach an emotional threshold for fabric eating.

Yet sixty-one per cent of cats in the survey didn't exhibit other continuing infantile behaviour patterns, suggesting perhaps that while we may encourage the development of wool eating, the underlying disposition is unaffected by our relationship with the cat.

Seventy-eight per cent of cats surveyed started to eat fabric during adolescence at 4-12 months. Twenty-two per cent began in response to some form of stress or trauma, such as medical illness, moving house or the acquisition of another cat, so perhaps the notion of the phenomenon being an inherited disposition requiring an environmental or emotional trigger is deserving of further investigation. It is apparent that once started on wool, most move on to other fabrics. Males seem just as likely to start as females but neutering only alleviated the behaviour in about 15% of cases. Some cats cease eating fabric altogether at about two years of age, perhaps because they learn to cope with the challenge or change that triggered the behaviour in the first place. A few are treatable, for instance those who have only been slightly pushed over the reaction threshold, though quite why some should respond and not others, is a mystery.

My first approach is to consider the prospect of over attachment being an important influence when its withdrawal or lack of constant availability of the owners leads to a form of stress-avoiding fabric eating. Such cases usually concern cats which are are only found to eat wool or fabric in their owners' absence, but which are often clingy and dependent when they are around. Treatment for them is encouragement to grow up, and to exchange any continuing infantile reactions with the owners for more adult ones and to keep affection in short doses. Wherever possible these cats are encouraged to go outside to further their level of stimulation, reduce the importance of owners and home for activity and help establish a less dependent relationship with the owners. Many appear to resolve, though how many are finding fabric sources in the homes of neighbours or from their washing lines is difficult to say.

Access to edible fabric must be made impossible for all fabric eaters, and sometimes denial for a few weeks causes the behaviour to cease. Direct negative conditioning by ambushing the cat with a water pistol while in mid chew has helped some, though often such tactics only produce a secret fabric eater. Indirect tactics using taste deterrents applied to specially laid towel baits can have a dramatic effect in deterring the cat from eating all fabric for evermore, though choice of deterrent is crucial. The traditional tastes of pepper, mustard and chilli or curry paste are invariably useless and simply broaden the cat's desire for a more exotic diet. Aromatic compounds such as menthol and oil of eucalyptus seem to be more successful and have reformed a number of ardent fabric eaters.

The best hope seems to lie with dietary management, though why this should be is difficult to explain when most fabric eaters have a perfectly normal healthy intake and their appetite for it is undiminished by having a stomach full of nylon sweater or woollen scarf. Providing a constantly available source of dry cat food, in addition to offering the usual diet, can redirect the desire to eat fabric onto more nutritional targets, and apparently without risk of weight gain. Most cats simply snack all day on the dry food and cut down voluntarily on the intake at usual mealtimes. Sometimes it helps to cease mealtimes altogether and simply leave a constant supply of dry food for the cat.

If keeping the stomach constantly active and partially full switches off the need to eat fabric in some sufferers, it may be due to the pleasurable comforting feeling of having food in the stomach rather than as an alternative form of intake. This may explain why adding fibre, which helps pad out the volume and passage time of traditional canned diets, can help prevent fabric eating. Padding such as bran can be added to the usual diet up to a point, but most will not accept too much. Instead, adding small lengths of finely chopped undyed wool or tissue to the diet may be more acceptable to the cat. This is a form of giving in, though can be a lot less expensive than letting the cat select his own fabric from the wardrobe. Other owners have resolved the problem of indiscriminate fabric eating by providing the cat with a towel to chew at dinner time. The cat takes a few mouthfuls of food, then chews and eats a portion of towel - a truly bizarre spectacle, but extremely effective in some cases.

With some fabric eaters the time taken to ingest food seems to be relevant to the frequency and severity of their fabric eating at other times. In the wild, cats would have to stalk, capture, handle and kill their prey before consuming it. Consumption itself would take some time as the cat would have to rip through the fur or feathers to expose the flesh. Such gory necessities are not precursors to eating for the pet cat. We hunt its food for it at the supermarket or pet shop, wrestle it from the cupboard, render it dead by piercing the box or cutting its lid off with a can opener and exposing all the good portions on a plate ready for the cat to consume at his leisure. No wonder so many cats have bad teeth - they never get used. If through the evolutionary development the cat has come to be so programmed as a hunter as to require the capture and prey element as a precursor to stimulating appetite or correct digestion, then the cat will benefit by being forced to invest more time in processing its food. This certainly seems to be the case with some fabric eaters which are forced to gain their food by spending much time chewing at the gristly meat and indigestible sinew attached to large bones rather than being offered free meals on a plate. Providing tough chunks of meat forces the cat to invest more time in its total feeding behaviour and helps reduce or alleviate entirely the desire to eat fabric at other times.

More than this I do not have to offer the owners of wool eating cats at present, though the results of the survey and details of the cats' pedigrees, where appropriate, are currently being processed further by fellow behaviourist Dr John Bradshaw and his student Diana Sawyer at the University of Southampton. Hopefully this will shed a little more light on to the origins of this strange phenomenon. We hope then to carry out more direct observational and analytical trials, especially on the cat's dietary and breeding aspects, and on the influences of stress on fabric eating, in collaboration with Dr Tim Gruffydd-Jones at the Dept of Veterinary Medicine at Bristol where I hold my monthly referral clinics. Tim is one of the leading authorities on feline medicine in the country and so with all angles covered we hope to enlarge our understanding of fabric eating in cats quite markedly over the coming year.

Pica – electric cables

Dear Mr Neville,

My Burmese cat, Rangoon, has a very dangerous behaviour problem. He loves to chew electric cables, but thankfully so far has yet to connect to the wires inside. I've tried telling him off but he seems incurably fascinated by them. Is there anything to do or is he doomed to be electrocuted one day when I'm caught napping?

Yours sincerely

Siobhan O'Flaherty

This apparently self-limiting behaviour could obviously be fatal to the cat, and also to the owner and her property because of the risk of fire. Understanding not one jot of the motivation to chew electric cables I took the view that an early interest in a movable play cable had developed into predatory handling, the end sequence of which is to consume the dead prey. So while some cats simply hunt mice, process them ready for eating and then lose interest prior to ingestion, Rangoon had jumped from stalking to consumption and learned to miss out the handling sequences. If it was a learned behaviour pattern perhaps a little negative conditioning could undo it, along with the offering of suitable alternative interests. Just for good measure we began by treating Rangoon as a fabric eater and altered his diet and feeding patterns in the hope of switching off any predatory or appetite motivated aspects of his behaviour. We tried to redirect his play and mock predatory activity by encouraging him to spend more time outdoors and with an attractive range of toys, some laced with catnip which fortunately proved irresistible to Rangoon. Then we allowed unsupervised access to electric cables suitably coated with eucalyptus oil and unplugged from the mains. Something from this battery of suggestions was successful as Rangoon no longer chews electric cables and leads a happy safe life. Friend and cartoonist Russell Jones decided that Rangoon was actually trying to administer electro-convulsive therapy to himself for some other unseen psychological disorder, and in the absence of any other serious explanation for this behaviour (which I have only encountered in two cats, both Burmese) I almost have to agree with him.

Pica – rubber

A good number of the owners who responded to my survey on fabric eating also mentioned that their cat ate rubber. Most restricted themselves to the delights of rubber bands with not a little risk of gastric problems. However, Wilberforce specialized in condoms. His career of lurking in the bedroom came to an abrupt end when the night after his owner had overcome his embarrassment and called me, Wilberforce made the mistake of trying to capture his prize after it had been fitted but not yet employed for its intended function. This was the last straw for Wilberforce's owner who, doubtless smarting in pain, took Wilberforce to the rescue shelter the next day. Quite why some cats, particularly Oriental breeds again, should be so obsessed by rubber I do not know, though they are often attracted by plastic as well. Treat as for electric cable eaters and pray!

Plant eating

Most cats eat more plant material than we realize probably in an effort to obtain a quickly digestible source of vitamins and minerals and roughage. Some regurgitate the grass with a portion of their dinner and this is believed to be a natural method of worming or helping to eject hairballs. Leopards can live for many weeks on plant material alone without ill effect, though all cats are obligate carnivores and cannot be maintained on a vegetarian diet for very long. Indoor cats should be provided with a tub of seedling sprouts to munch on, so as to discourage consumption of houseplants. These are available from pet shops and being more attractive than most houseplants are readily consumed in preference. A little negative conditioning with a water pistol can help convince the cat obsessed with eating one's favourite plant, but this is difficult when the plant in question lives in someone else's garden.

However, Suki caused great concern because she had to undergo surgery six times to remove a jagged edge of a leaf of one particular plant from her throat. Identification and preferably removal of the plant is the obvious course of action to take, but this has so far proved impossible. The cat seems to have an irresistible urge to eat only that plant, though we are presently hoping that a few weeks' confinement indoors after the last operation and the opportunity to eat seedling sprouts and a dry diet, may help redirect the cat's desires and alter his habits. If we could only take an identifiable sample of the plant, we may find that like catnip, it contains some irresistible compound which could then be offered to the cat in a safer form.

Other interesting cases have concerned cats with passions for eating certain foods, or simply all food. Polyphagia, as this is known, is not at all common in cats as they are renowned for limiting their intake of food to suit their needs. One does encounter the occasional feline barrel however, and aside from the many medical conditions which can cause an increase in appetite, this is most frequently observed in permanently indoor cats. Boredom eating enables the cat to occupy itself, and as with humans, eating seems to help relieve stress. Treatment usually involves providing more carefully controlled portion feeding and denying access to food at other times, coupled with greater stimulation and activation of the unoccupied cat. Outside walks and acquiring a second cat can help dramatically in the treatment of boredom eating.

An alcoholic

Quite what to do about a client's cat with a passion for Campari and soda was a little more difficult. Gradual withdrawal of access to alcohol was advised as a slow drying out programme for Slosher, who had been used to a daily drink for over three years. I also advised that the soda level be raised at the expense of the Campari and hoped that cold turkey would not set in.

Sexual problems

And so finally to sex. Sadly this is a short and disappointing section as cats hardly ever present sexual problems and are no match whatsoever for the often randy and excessive exploits of the young male dog. Females are far less likely to show unwanted or bizarre sexual behaviour in cats or dogs, and the fate of most male cats is to be castrated at an early age. On entering adolescence and perhaps just prior to castration (or even the cause of it) toms may occasionally mount their owners arm, cushions or furniture and masturbate, particularly following a period of stroking or excitement, but the behaviour is usually short lived and bears none of the insistence of the randy young terrier's assault on one's leg.

Nor are cats famous as flashers like so many dogs, including my own Colonel, the Dobermann, whose earlier name prior to the operation was 'Pencil'. Indeed a cat's sexual responses seem unexpressed except under the single stimulus of the smell of a queen in season. The entire male can detect this from miles away and will pursue with the utmost vigour. But courtship is usually short, and would be looked upon as rape were it not for the willingness of the queen to adopt lordosis, the receptive mating position. The performance is usually very quick though may be repeated a number of times despite the fact that withdrawal is a painful affair for the queen, due to the backwards facing barbs on the tom's penis. This stimulation of the walls of the female's vagina, while painful, stimulate ovulation, so ensuring that every mating has the best possible chance of success. It also means that kittens in a litter may have different fathers if several suitors are at the scene when the queen is receptive.

A cat's sex life is predictable and neat, and I have only ever dealt with two cases of misdirected sexual behaviour. One concerned a totally inexplicable case of a three-year-old neutered male cat who persistently mounted one particular radiator knob. It remains high on my list of favourite self-resolving problems though I am none the wiser as to its origins. High on my list of treated problems is also:

Dear Mr Neville,

Over the last few days, Thumper, my neutered ginger tom has fallen in love with Jasmine, my eight year old Dachshund. He follows her everywhere, yowling and sniffing and tries to mount her at every opportunity. She is clearly distressed and tries to hide, but to no avail. He is like a cat possessed and has kept up his advances almost continually. What on earth is happening?

Your sincerely

Richard Calvin

As it turned out poor Thumper was responding quite normally to a series of unusual pheremones being given off by Jasmine as a result of an internal tumour. These triggered his sexual response in a similar manner as do the smells of a queen calling, even though he was neutered. The reception areas of the brain remain unaltered by castration and ordinarily the normal responses are prevented in the absence of male hormones. Occasionally other hormones can compensate for the lack of androgens and engender sexual responses. While it is not uncommon to see castrated dogs continuing to show sexual responses after castration, it is very rare in cats. Thumper was probably one such exception and could do nothing else but respond to the plainly enticing smells of Jasmine. The vet treated the dog and the cat's behaviour problems disappeared overnight.

Epilogue

If this journey along the roads of feline insanity and undesirable behaviour has helped even one owner enjoy and understand their relationship with their cat a little more, then its aim will have been achieved. My own cats Scribble and Bullet have joined me from time to time during the writing of this book to gaze down at me from the top of my word processor and form their own opinions as to why their previously doting mother-figure owner should abandon them for the sake of tapping away on a keyboard for hours on end. Perhaps they will produce the sequel to this book and call it *Do Owners Need Shrinks?* The answer to that, and to the title of this book *Do Cats Need Shrinks?* is YES! Not all the time perhaps, but occasionally and definitely, YES!

Help For Cats with Behaviour Problems

My cases are seen strictly on referral from practising veterinary surgeons, so if your cat is presenting a behaviour problem, please seek their advice first. If they would like to refer you, I could see you and your cat at the most convenient of the following monthly venues, or can occasionally manage house-calls depending on where you live and where I am.

Serving the South West, Wales and the Midlands:

The Feline Behaviour Referral Clinic
Department of Veterinary Medicine
University of Bristol
Langford House
Langford
Bristol BS18 7DU

Serving London and the South East:

The Feline Behaviour Clinic
Woodthorpe Veterinary Group
6 Woodthorpe Road
Putney
London SW15 6UQ

Serving the North West:

The Animal Behaviour Clinic (for dogs and cats)
Rutland House Veterinary Hospital
Cowley Hill Lane
St Helens
Lancashire WA10 2AW

Serving Dublin and surrounding regions (for dogs and cats)

Mssrs Wilson and Kelly MsRCVS
The Lodge
Old Conna Avenue
Dublin Road
Bray
Eire

Association of Pet Behaviour Consultants

My practice is one of the original members of the Association of Pet
Behaviour Consultants (APBC). The APBC comprises five professional
member practices nationwide, so if your pet has a behaviour problem,
a member of the Association may be able to help you in your own area.
For details of my own or the nearest member's practice please write
to the Honorary Secretary, APBC, 50 Pall Mall, London SW1.
I'm sure we can help!